Association for Women in Mathematics Series

Volume 13

Series Editor
Kristin Lauter
Microsoft Research
Redmond, Washington, USA

Association for Women in Mathematics Series

More information about this series at http://www.springer.com/series/13764

Erin Wolf Chambers • Brittany Terese Fasy
Lori Ziegelmeier

Editors

Research in Computational Topology

Editors
Erin Wolf Chambers
Department of Computer Science
St Louis University
St Louis, MO, USA

Brittany Terese Fasy
Gianforte School of Computing, and
Department of Mathematical Sciences
Montana State University
Bozeman, MT, USA

Lori Ziegelmeier
Department of Mathematics,
Statistics, & Computer Science
Macalester College
Saint Paul, MN, USA

ISSN 2364-5733 ISSN 2364-5741 (electronic)
Association for Women in Mathematics Series
ISBN 978-3-030-07810-2 ISBN 978-3-319-89593-2 (eBook)
https://doi.org/10.1007/978-3-319-89593-2

This Springer imprint is published by the registered company Springer International Publishing AG part of Springer Nature.
The registered company address is: Gewerbestrasse 11, 6330 Cham, Switzerland

Preface

In August 2016, the first Workshop for Women in Computational Topology was held at the Institute for Mathematics and its Applications (IMA) in Minneapolis, MN, with generous sponsorship from the National Science Foundation and Microsoft Research. A group of 27 women, ranging from undergraduates to full professors, spent one week working in groups on cutting-edge problems from the field; we were also joined by five children of the participants, who joined for various social events throughout the week. In addition to the working groups, the workshop also included a poster session, a panel on work/life balance, and many opportunities for networking (Fig. 1).

Fig. 1 The women in attendance at WinCompTop 2016

At the conclusion of the workshop, each working group compiled the results of their research, which became the core of this volume. We also solicited submissions on related topics from the Women in Computational Topology listserv community, which includes both women and their allies. Submissions went through a two-phase single blind review process.

A Brief Introduction to Computational Topology

Computational topology is a new and exciting field at the intersection of mathematics, statistics, and computer science, bringing experts from algebra, topology, algorithms, and applications areas together. Core areas in this research field include topological data analysis and persistent homology, surface reconstruction, planar and surface embedded graphs and algorithms on such objects, and three-manifold recognition. Techniques and algorithms are useful in a range of application areas, including graphics, medical imaging, materials engineering, and GIS (Fig. 2, 3, 4, 5, and 6). The papers in this volume are self-contained; however, the following textbooks are great resources [1–3].

Fig. 2 WinCompTop 2016 conference organizers and editors of this volume Lori Ziegelmeier, Erin Wolf Chambers, and Brittany Terese Fasy

Fig. 3 Banquet at the Tea House

Fig. 4 Bei Wang describing her career path during the panel on work-life balance

Fig. 5 Making progress in Yusu Wang's working group

Working Groups

Our workshop had four working groups that represented a range of the research areas in computational topology, with each group's topic chosen by a leader in the field. Participants were able to read a brief overview of the problem and rate their interest, and then the organizers arranged participants into four roughly equal groups according to these preferences. These four groups are briefly described here:

Yusu Wang (The Ohio State University) led a group that explored a new method of persistence-based profiles to compare metric graphs, examining what information is captured by these persistence-based profiles and understanding their discriminative power. This working group's paper is the third chapter in the volume.

The next working group's paper follows the paper by Wang's group. Carola Wenk (Tulane University) and Brittany Terese Fasy (Montana State University) considered the use of topology in map construction and comparison, particularly understanding directed graphs with multiple lanes and overpasses.

Nina Amenta (University of California, Davis) led a group which examined the problem of producing an explicit representation of a surface S from an input point cloud P assumed to lie on or near S. This working group's paper is the seventh chapter in the volume.

Fig. 6 Making progress in Carola Wenk and Brittany Terese Fasy's working group

Finally, Giseon Heo (University of Alberta) led a project that extended one dimensional scale-space persistent homology (a fundamental tool in computational topology) to a pseudo-multidimensional persistence tool that can be applied to a variety of applications. This working group's paper is the last chapter in the volume.

Contributed Papers

At the conclusion of the workshop, we solicited papers from the working groups as well as from the broader women in computational topology network. This volume represents the ten accepted contributions from that pool. The volume begins with more theoretical results and then transitions to more application-driven papers.

The volume begins with a result by Landi that proves a lower bound on the interleaving distance of persistence modules using matching distance of rank invariants, which offers an alternate proof of the stability of rank invariants.

In the next chapter, Chambers and Letscher consider generalizing persistent homology to spaces where the underlying inclusions form a directed acyclic graph, and give an algorithm to compute the groups for all subgraphs with a common source and sink vertex.

Continuing with the theme of graphs, in the first working group paper, Gasparovic et al. completely classify the one-dimensional persistence diagram of the intrinsic Čech complex of metric graphs. In particular, they prove that no features die before an appropriate filtration value and no other homology classes are created due to interference among loops in the metric graph.

Another article from a working group is presented in the next chapter. In this article, Bittner et al. investigate comparisons of graphs and data on graphs. In particular, they define distance measures between (embedded, directed) graphs, as well as distance measures between data or information on graphs (which may or may not be the same underlying graph).

In the next chapter, Adams et al. consider the problem of sweeping a planar region with a continuous curve, while keeping the length of the curve as small as possible, so that no intruder can avoid the curve. They then discuss analytic bounds on such a curve in terms of the geodesic Fréchet distance.

Bendich et al. propose a flexible, multiscale approach to organize, visualize, and understand point cloud data sampled near stratified spaces by introducing an algorithm that produces a cover tree reflecting the local geometry of the space and then uncovering the strata making up this underlying space.

In the next working group chapter, Amenta et al. study the Delaunay triangulation of an (ϵ, δ)-sample P of $\mathbb{R}^2$, and show that the set of *anchored* Delaunay triangles (the triangles containing their Voronoi center) are dense.

The last three chapters in the volume are more application-driven, applying the topological technique of persistent homology to a variety of data sets.

Banman and Ziegelmeier explore the topological characteristics of country development and geography using indicators such as gross domestic product per capita, average life expectancy, etc., revealing similarities and localized development patterns of countries at multiple scales.

Wubie et al. use persistent homology as a tool to detect patterns in data that go beyond traditional clustering measures, using a data set generated from the Scientific Registry of Transplant Recipients. Their analysis is able to better discriminate between the identified clusters in some of the tests, when compared to other clustering methodologies.

The volume concludes with the final working group project. Betancourt et al. explore a variety of data sets—such as craniofacial shape, Lissajous knots, and the Kuramoto-Sivashinsky partial differential equation—that vary across two inputs, the filtration parameter and another such as curvature or time. They analyze these data sets using the heat map pseudo-multidimensional persistence technique of Xia and Wei [4] which provides further insight into these data sets.

Contents

The Rank Invariant Stability via Interleavings

Claudia Landi

Abstract A lower bound for the interleaving distance on persistence modules is given in terms of matching distance of rank invariants. This offers an alternative proof of the stability of rank invariants. As a further contribution, also the internal stability of the rank invariant is proved in terms of interleavings.

1 Introduction

Increasingly often in recent years, the shape of data has been analyzed using persistent homology [7]. Given a topological space (e.g., a simplicial complex built upon a finite set of points in $\mathbb{R}^n$), one constructs a nested family of subspaces called a filtration and studies the topological events occurring along the filtration. These are encoded in a structure known as a persistence module consisting of a family of vector spaces connected by linear maps. In fact, a persistence module is usually obtained from a filtration by considering the vector spaces and the linear maps induced in homology by the inclusion maps of the filtration.

The theory of persistence modules is completely understood when filtrations only depend on one parameter [14], whereas still present many open problems in the multiparameter case, commonly referred to as the multidimensional persistence [8]. On the other hand, applications strongly motivate the interest in multi-filtrations.

Recently, the problem of comparing persistence modules in a stable and optimal way has been successfully solved for any number of parameters by using families of natural transformations between persistence modules, known as interleavings [13]. The interleaving distance measures the amount of shift necessary to map two persistence modules into each other. In concrete cases, computing the interleaving distance is still not a viable option because of its complexity, let alone the complexity of computing the persistence modules themselves [9]. This suggests that it may be useful to find estimates for it.

C. Landi (✉)
Dipartimento di Scienze e Metodi dell'Ingegneria, Università di Modena e Reggio Emilia, Reggio Emilia, Italy
e-mail: claudia.landi@unimore.it

© The Author(s) and the Association for Women in Mathematics 2018
E. W. Chambers et al. (eds.), *Research in Computational Topology*, Association for Women in Mathematics Series 13, https://doi.org/10.1007/978-3-319-89593-2_1

The rank invariant [8], also known in the literature as persistent Betti numbers (PBNs) [11] or, for zeroth homology, as a size function [3], is the most studied invariant of persistence modules. The rank invariant only captures the rank values of the linear maps defining the persistence module. Thus, it only gives a summary of the persistence module, but often this is sufficient for applications (see, e.g.,[1, 4, 6, 15]).

The primary goal of this paper is to show that the multidimensional matching distance on rank invariants proposed in [11] provides a lower bound for the interleaving distance. From a different standpoint, this fact can be viewed as a new proof that the rank invariant is a stable invariant when rank invariants are compared via one-dimensional reductions along lines. Indeed, the secondary goal of this paper is to obtain a new proof of the rank invariant stability, that is the property of undergoing small changes when the input data is slightly perturbed. In the present paper, the stability of the rank invariant is proved leveraging interleavings, whereas one previous proof presented in [11] was based on the explicit construction of a family of one-parameter filtrations behaving as the given multiparameter filtration, and another proof proposed in [10] was by construction of an analogue of the persistence diagram for PBNs, called a persistence space. It has to be underlined that the stability proofs of [11] and [10] do not seem to yield estimates for the interleaving distance.

As a final contribution of the paper, we prove the internal stability of the rank invariant, that is the property of undergoing small changes when parameters are slightly perturbed. Internal stability is a key property for the multidimensional matching distance computation [5]. Also in this case, a previous proof of this property already exists (cf. [11]), but the new proof presented here is based solely on interleavings.

The paper is organized as follows. In Sect. 2, the definitions of persistence module and rank invariant are reviewed together with the relevant metrics: interleaving distance, bottleneck distance, and matching distance. Section 3 contains the proof that the multidimensional matching distance on rank invariants provides a lower bound for the interleaving distance on pointwise finite dimensional persistence modules (Theorem 1). As a consequence, we get yet another proof of the stability of rank invariants (Corollary 1). The proof of the internal stability of the rank invariant (Theorem 2) is given in Sect. 4. A brief discussion concludes the paper.

2 Background Definitions

Let $n \in \mathbb{N}_{>0}$. For every $u = (u_i), v = (v_i) \in \mathbb{R}^n$, we write $u \preceq v$ (resp. $u \prec v$) if and only if $u_i \leq v_i$ (resp. $u_i < v_i$) for $i = 1, \ldots, n$. Note that $v \prec u$ is not the negation of $u \preceq v$. When $\mathbb{R}^n$ is viewed as a vector space, its elements are denoted in bold font. Moreover, in this case, we endow $\mathbb{R}^n$ with the max norm defined by $\|\mathbf{v}\|_\infty = \max_i |v_i|$.

For a field $\mathbb{F}$, an *n-dimensional persistence module* $\mathbf{M}$ is a family $\{\mathbf{M}_u\}_{u \in \mathbb{R}^n}$ of $\mathbb{F}$-vector spaces, together with a family of linear maps $\{\varphi_\mathbf{M}(u, v) : \mathbf{M}_u \to \mathbf{M}_v\}_{u \preceq v}$ such that, for all $u \preceq v \preceq w \in \mathbb{R}^n$, $\varphi_\mathbf{M}(u, u) = \mathrm{Id}_{\mathbf{M}_u}$, and $\varphi_\mathbf{M}(v, w) \circ \varphi_\mathbf{M}(u, v) = \varphi_\mathbf{M}(u, w)$. We call the maps $\varphi_\mathbf{M}(u, v)$ *transition maps*. $\mathbf{M}$ is said to be *pointwise finite dimensional*, if $\dim(\mathbf{M}_u) < \infty$ for all $u \in \mathbb{R}^n$.

A morphism $\alpha : \mathbf{M} \to \mathbf{N}$ of persistence modules is a collection of linear maps $\alpha(u) : \mathbf{M}_u \to \mathbf{N}_u$ such that $\alpha(v) \circ \varphi_\mathbf{M}(u, v) = \varphi_\mathbf{N}(u, v) \circ \alpha(u)$, for all $u \preceq v \in \mathbb{R}^n$.

For $\mathbf{M}$ a persistence module, and for $\varepsilon \geq 0$ a real value, $\mathbf{M}(\varepsilon)$ denotes the module $\mathbf{M}$ diagonally shifted by $\boldsymbol{\varepsilon} = (\varepsilon, \varepsilon, \ldots, \varepsilon)$: $\mathbf{M}(\boldsymbol{\varepsilon})_u = \mathbf{M}_{u + \boldsymbol{\varepsilon}}$, and for $u \preceq v \in \mathbb{R}^n$, $\varphi_{\mathbf{M}(\boldsymbol{\varepsilon})}(u, v) = \varphi_\mathbf{M}(u + \boldsymbol{\varepsilon}, v + \boldsymbol{\varepsilon})$. We also let $\varphi_\mathbf{M}(\boldsymbol{\varepsilon}) : \mathbf{M} \to \mathbf{M}(\boldsymbol{\varepsilon})$ be the diagonal ε-transition morphism, that is the morphism whose restriction to $\mathbf{M}_u$ is the linear map $\varphi_\mathbf{M}(u, u + \boldsymbol{\varepsilon})$ for all $u \in \mathbb{R}^n$. We say that two *n*-modules $\mathbf{M}$ and $\mathbf{N}$ are ε-interleaved if there exist morphisms $\alpha : \mathbf{M} \to \mathbf{N}(\varepsilon)$ and $\beta : \mathbf{N} \to \mathbf{M}(\varepsilon)$ such that $\beta(\boldsymbol{\varepsilon}) \circ \alpha = \varphi_\mathbf{M}(2\boldsymbol{\varepsilon})$ and $\alpha(\boldsymbol{\varepsilon}) \circ \beta = \varphi_\mathbf{N}(2\boldsymbol{\varepsilon})$. The *interleaving distance* on persistence modules is defined by setting

$$d_I(\mathbf{M}, \mathbf{N}) = \inf\{\varepsilon \in [0, +\infty) : \mathbf{M} \text{ and } \mathbf{N} \text{ are } \varepsilon\text{-interleaved}\}$$

with the convention that the infimum over the empty set is $+\infty$.

The theory of persistence modules is well understood in dimension $n = 1$, and nicely reviewed in [2] (see also [14]). In particular, for $n = 1$, any pointwise finite dimensional persistence module is completely representable by a unique multiset of intervals $B(\mathbf{M})$, called a *barcode*, or, equivalently, by a multiset of points of $\mathbb{R}^2$, called a *persistence diagram*. The latter is obtained from a barcode by identifying an interval with endpoints a, b to a point in $\mathbb{R}^2$ with coordinates (a, b). The *bottleneck distance* d_B is an easily computable pseudo-metric on barcodes defined as the smallest possible cost of a partial matching M between the intervals in the barcode $B(\mathbf{M})$ and those in $B(\mathbf{N})$:

$$d_B(B(\mathbf{M}), B(\mathbf{N})) = \inf_M c(M).$$

Here, the cost $c(M)$ of a partial matching M is the greatest among the costs of each matched pair of intervals (taken to be equal to the ℓ^∞ distance of the corresponding pair of points in $\mathbb{R}^2$) and the costs of each unmatched interval (taken to be equal to $|a - b|/2$ for an interval with endpoints a and b). The Algebraic Stability of Persistence Barcodes states that, for any two pointwise finite dimensional persistence modules $\mathbf{M}$ and $\mathbf{N}$ of dimension 1, it holds that $d_B(B(\mathbf{M}), B(\mathbf{N})) \leq d_I(\mathbf{M}, \mathbf{N})$.

In dimension $n > 1$, the persistence module structure is still matter of investigation, and numeric invariants are often used instead. The *rank invariant* of persistence modules is defined by setting $\rho_\mathbf{M}(u, v) = \mathrm{rank}\varphi_\mathbf{M}(u, v)$ for $u \preceq v \in \mathbb{R}^n$. A persistence module is called q-tame if $\rho_\mathbf{M}(u, v) < \infty$ for any $u \prec v$. Pointwise finite dimensional persistence modules are q-tame. In [11], a readily computable pseudo-metric on rank invariants of q-tame persistence modules, the multidimensional matching distance d_{match}, is defined via one-dimensional reductions:

$$d_{match}(\rho_{\mathbf{M}}, \rho_{\mathbf{N}}) = \sup_L \hat{m} \cdot d_B(B(\mathbf{M}_L), B(\mathbf{N}_L))$$

where L varies in the set of all the lines of $\mathbb{R}^n$ whose direction $\mathbf{m} = (m_i)$ is such that $\hat{m} = \min_i m_i$ is strictly positive, and $\mathbf{M}_L$, $\mathbf{N}_L$ denote the restrictions of $\mathbf{M}$, $\mathbf{N}$ to the line L. For rank invariants of persistence modules built from sublevel sets of continuous functions on triangulable spaces, the multidimensional matching distance is a metric, thus implying that it captures the full structure of such invariants.

In the next section, the connections between the rank invariant and interleavings are highlighted.

3 A Lower Bound for the Interleaving Distance

Given a line L in $\mathbb{R}^n$ parameterized by $u = s\mathbf{m} + b$, with $\hat{m} = \min_i m_i > 0$, we denote by $\mathbf{M}_L$ the persistence module parameterized by $s \in \mathbb{R}$ and obtained by restriction of $\mathbf{M}$ to L: $(\mathbf{M}_L)_s = \mathbf{M}_u$ and $\varphi_{\mathbf{M}_L}(s, s') = \varphi_{\mathbf{M}}(u, u')$ for $u = s\mathbf{m} + b$ and $u' = s'\mathbf{m} + b$.

Lemma 1 *If $\mathbf{M}$ and $\mathbf{N}$ are ε-interleaved, then $\mathbf{M}_L$ and $\mathbf{N}_L$ are $\frac{\varepsilon}{\hat{m}}$-interleaved.*

Proof Because $\mathbf{M}$ and $\mathbf{N}$ are ε-interleaved, there exist two morphisms $\alpha : \mathbf{M} \to \mathbf{N}(\varepsilon)$ and $\beta : \mathbf{N} \to \mathbf{M}(\varepsilon)$ such that $\beta(u + \varepsilon) \circ \alpha(u) = \varphi_{\mathbf{M}}(u, u + 2\varepsilon)$ and $\alpha(u + \varepsilon) \circ \beta(u) = \varphi_{\mathbf{N}}(u, u + 2\varepsilon)$, for every $u \in \mathbb{R}^n$.

We need to prove that there are two morphisms $\alpha_L : \mathbf{M}_L \to \mathbf{N}_L(\frac{\varepsilon}{\hat{m}})$ and $\beta_L : \mathbf{N}_L \to \mathbf{M}_L(\frac{\varepsilon}{\hat{m}})$ such that $\beta_L\left(s + \frac{\varepsilon}{\hat{m}}\right) \circ \alpha_L(s) = \varphi_{\mathbf{M}_L}(s, s + 2\frac{\varepsilon}{\hat{m}})$ and $\alpha_L(s + \frac{\varepsilon}{\hat{m}}) \circ \beta_L(s) = \varphi_{\mathbf{N}_L}(s, s + 2\frac{\varepsilon}{\hat{m}})$, for every $s \in \mathbb{R}$. The idea underlying the construction of α_L and β_L is illustrated in Fig. 1.

For $s \in \mathbb{R}$, take the points $u(s) = s\mathbf{m} + b$, $u'(s) = \left(s + \frac{\varepsilon}{\hat{m}}\right)\mathbf{m} + b$, and $u''(s) = \left(s + 2\frac{\varepsilon}{\hat{m}}\right)\mathbf{m} + b$ on L. Setting $\hat{\imath} = \arg\min_i m_i$, a direct computation shows that $u'(s)$ is the point of L such that $u'_{\hat{\imath}}(s) = u_{\hat{\imath}}(s) + \varepsilon$. Analogously, $u''(s)$ turns out to be the point of L such that $u''_{\hat{\imath}}(s) = u'_{\hat{\imath}}(s) + \varepsilon$.

We are now ready to construct α_L and β_L. Recall that $u, u', u'' \in L$ are functions of s. We define $\alpha_L : \mathbf{M}_L \to \mathbf{N}_L(\frac{\varepsilon}{\hat{m}})$ by setting $\alpha_L(s) = \alpha(u' - \varepsilon) \circ \varphi_{\mathbf{M}}(u, u' - \varepsilon)$ (red arrows in Fig. 1). Analogously, we define $\beta_L : \mathbf{N}_L \to \mathbf{M}_L(\frac{\varepsilon}{\hat{m}})$ by setting $\beta_L(s) = \beta(u' - \varepsilon) \circ \varphi_{\mathbf{N}}(u, u' - \varepsilon)$ (green arrows in Fig. 1). α_L and β_L are well defined because a direct computation shows that $u \preceq u' - \varepsilon$.

Observing that $\beta_L(s) = \varphi_{\mathbf{M}}(u + \varepsilon, u') \circ \beta(u)$ by definition of morphism of persistence modules, we get

$$\beta_L\left(s + \frac{\varepsilon}{\hat{m}}\right) \circ \alpha_L(s) = \varphi_{\mathbf{M}}(u' + \varepsilon, u'') \circ \beta(u') \circ \alpha(u' - \varepsilon) \circ \varphi_{\mathbf{M}}(u, u' - \varepsilon).$$

Fig. 1 The maps α_L and β_L defined in Lemma 1: α_L is given by composition of the red maps, β_L is given by composition of the green maps. The diamond diagrams commute

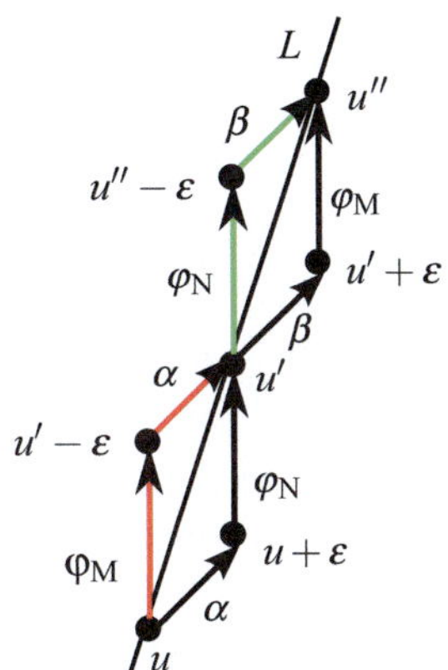

Recalling that $\beta(u') \circ \alpha(u' - \boldsymbol{\varepsilon}) = \varphi_{\mathbf{M}}(u' - \boldsymbol{\varepsilon}, u' + \boldsymbol{\varepsilon})$, we obtain

$$\beta_L \left(s + \frac{\varepsilon}{\hat{m}}\right) \circ \alpha_L(s) = \varphi_{\mathbf{M}}(u' + \boldsymbol{\varepsilon}, u'') \circ \varphi_{\mathbf{M}}(u' - \boldsymbol{\varepsilon}, u' + \boldsymbol{\varepsilon}) \circ \varphi_{\mathbf{M}}(u, u' - \boldsymbol{\varepsilon})$$

immediately yielding

$$\beta_L \left(s + \frac{\varepsilon}{\hat{m}}\right) \circ \alpha_L(s) = \varphi_{\mathbf{M}}(u, u''),$$

or equivalently

$$\beta_L \left(s + \frac{\varepsilon}{\hat{m}}\right) \circ \alpha_L(s) = \varphi_{\mathbf{M}_L} \left(s, s + 2\frac{\varepsilon}{\hat{m}}\right).$$

Analogously, observing that

$$\alpha(u'' - \boldsymbol{\varepsilon}) \circ \varphi_{\mathbf{M}}(u', u'' - \boldsymbol{\varepsilon}) = \varphi_{\mathbf{N}}(u' + \boldsymbol{\varepsilon}, u'') \circ \alpha(u')$$

by definition of morphism of persistence modules, we have

$$\alpha_L \left(s + \frac{\varepsilon}{\hat{m}}\right) \circ \beta_L(s) = \varphi_{\mathbf{N}}(u' + \boldsymbol{\varepsilon}, u'') \circ \alpha(u') \circ \beta(u' - \boldsymbol{\varepsilon}) \circ \varphi_{\mathbf{N}}(u, u' - \boldsymbol{\varepsilon}).$$

Hence, from $\alpha(u') \circ \beta(u' - \boldsymbol{\varepsilon}) = \varphi_{\mathbf{N}}(u' - \boldsymbol{\varepsilon}, u' + \boldsymbol{\varepsilon})$, we get

$$\alpha_L \left(s + \frac{\varepsilon}{\hat{m}}\right) \circ \beta_L(s) = \varphi_{\mathbf{N}}(u' + \boldsymbol{\varepsilon}, u'') \circ \varphi_{\mathbf{N}}(u' - \boldsymbol{\varepsilon}, u' + \boldsymbol{\varepsilon}) \circ \varphi_{\mathbf{N}}(u, u' - \boldsymbol{\varepsilon})$$

which immediately yields

$$\alpha_L \left(s + \frac{\varepsilon}{\hat{m}}\right) \circ \beta_L(s) = \varphi_{\mathbf{N}}(u, u'')$$

or equivalently

$$\alpha_L \left(s + \frac{\varepsilon}{\hat{m}}\right) \circ \beta_L(s) = \varphi_{\mathbf{N}_L}\left(s, s + 2\frac{\varepsilon}{\hat{m}}\right).$$

$\square$

Theorem 1 *For every pointwise finite dimensional modules $\mathbf{M}$ and $\mathbf{N}$, it holds that $d_{match}(\rho_{\mathbf{M}}, \rho_{\mathbf{N}}) \leq d_I(\mathbf{M}, \mathbf{N})$.*

Proof By definition $d_{match}(\rho_{\mathbf{M}}, \rho_{\mathbf{N}}) = \sup_{L:u=s\mathbf{m}+b} \hat{m} \cdot d_B(B(\mathbf{M}_L), B(\mathbf{N}_L))$ where L varies in the set of all the lines parameterized by $u = s\mathbf{m} + b$, with $\mathbf{m}$ such that $\hat{m} = \min_i m_i > 0$, $\|\mathbf{m}\|_\infty = 1$, and b such that $\sum_{i=1}^n b_i = 0$. By Lemma 1, for any such line L, if $\mathbf{M}$ and $\mathbf{N}$ are ε-interleaved then $\mathbf{M}_L$ and $\mathbf{N}_L$ are $\frac{\varepsilon}{\hat{m}}$-interleaved. Thus, $\hat{m} \cdot d_I(\mathbf{M}_L, \mathbf{N}_L) \leq d_I(\mathbf{M}, \mathbf{N})$, yielding $\hat{m} \cdot d_B(B(\mathbf{M}_L), B(\mathbf{N}_L)) \leq d_I(\mathbf{M}, \mathbf{N})$ by the Algebraic Stability of Persistence Barcodes. Hence the claim. $\square$

From the previous result, we immediately deduce the stability of rank invariants of persistence modules of sublevel set filtrations.

Corollary 1 *For $i \geq 0$, homeomorphic topological spaces X and Y, and continuous functions $\gamma_X : X \to \mathbb{R}^n$ and $\gamma_Y : Y \to \mathbb{R}^n$, assume that the persistence modules $H_i(\gamma_X)$ and $H_i(\gamma_Y)$ induced in homology by the sublevel set filtrations of γ_X and γ_Y, respectively, are pointwise finite dimensional. Then, we have*

$$d_{match}\left(\rho_{H_i(\gamma_X)}, \rho_{H_i(\gamma_Y)}\right) \leq \inf_{h \in \mathscr{H}} \sup_{p \in X} \|\gamma_X(p) - \gamma_Y \circ h(p)\|_\infty$$

where $\mathscr{H}$ is the set of homeomorphisms from X to Y.

Proof It is sufficient to apply Theorem 1, and recall that the interleaving distance enjoys the stability property as shown in [13]. $\square$

4 Internal Stability of the Rank Invariant

Let us now consider the problem of how the barcode $B(\mathbf{M}_L)$ changes as L changes, provided that L never gets parallel to the coordinate axes. The goal is to show that such variation is continuous with respect to the bottleneck distance.

For L parameterized by $u = s\mathbf{m} + b$, with $\hat{m} = \min_i m_i > 0$, and L' parameterized by $u = s\mathbf{m}' + b'$, with $\hat{m}' = \min_i m'_i > 0$, we denote by $\mathbf{M}_L$ and $\mathbf{M}_{L'}$ the persistence modules parameterized by $s \in \mathbb{R}$ obtained by restriction of $\mathbf{M}$ to L and L', respectively, and consider the interleavings between $\mathbf{M}_L$ and $\mathbf{M}_{L'}$.

Lemma 2 *Let $\mathbf{M}$ be a pointwise finite dimensional persistence module. Assume there exist $c \prec C \in \mathbb{R}^n$ such that $\varphi_{\mathbf{M}}(u, u')$ is an isomorphism for every $u, u' \in \mathbb{R}^n$ with $C \preceq u \preceq u'$ or $u \preceq u' \preceq c$. Then, for L and L' parameterized by $u = s\mathbf{m} + b$, with $\hat{m} = \min_i m_i > 0$, and $u = s\mathbf{m}' + b'$, with $\hat{m}' = \min_i m'_i > 0$, respectively, the persistence modules $\mathbf{M}_L$ and $\mathbf{M}_{L'}$ are η-interleaved with*

$$\eta = \frac{\left(\max\left\{\|c\|_\infty, \|C\|_\infty\right\} + \max\left\{\|b\|_\infty, \|b'\|_\infty\right\}\right) \cdot \|\mathbf{m} - \mathbf{m}'\|_\infty + \|b - b'\|_\infty}{\hat{m} \cdot \hat{m}'}.$$

Proof Let us construct morphisms $\alpha : \mathbf{M}_L \to \mathbf{M}_{L'}(\eta)$ and $\beta : \mathbf{M}_{L'} \to \mathbf{M}_L(\eta)$ such that $\beta(s + \eta) \circ \alpha(s) = \varphi_{\mathbf{M}_L}(s, s + 2\eta)$ and $\alpha(s + \eta) \circ \beta(s) = \varphi_{\mathbf{M}_{L'}}(s, s + 2\eta)$, for every $s \in \mathbb{R}$.

In order to construct α, we preliminarily consider the projection maps $\pi_L^\uparrow :$ $\mathbb{R}^n \to L$ and $\pi_L^\downarrow : \mathbb{R}^n \to L$ defined as follows. $\pi_L^\uparrow$ takes a point $u \in \mathbb{R}^n$ to the point $\bar{u} \in L$ such that $u \preceq \bar{u}$ and $\bar{u}_\iota = u_\iota$ for some ι with $1 \leq \iota \leq n$. $\pi_L^\downarrow$ takes a point $u \in \mathbb{R}^n$ to the point $\underline{u} \in L$ such that $\underline{u} \preceq u$ and $\underline{u}_\iota = u_\iota$ for some ι with $1 \leq \iota \leq n$. The points $\pi_L^\uparrow(u)$ and $\pi_L^\downarrow(u)$ are unique in L because $\hat{m} = \min_i m_i > 0$. Next, we consider the points $\bar{C} = \pi_L^\uparrow(C)$ and $\underline{c} = \pi_L^\downarrow(c)$ in L. It holds that $\bar{C} \preceq u$ implies $C \preceq u$, and $u \preceq \underline{c}$ implies $u \preceq c$.

Moreover, for $u \preceq c$, we set

$$\psi_{\mathbf{M}}(u, u') = \begin{cases} \varphi_{\mathbf{M}}(u, u') & \text{if } u \preceq u', \\ \varphi_{\mathbf{M}}(u', u)^{-1} & \text{if } u' \preceq u. \end{cases}$$

We note that $\varphi_{\mathbf{M}}(u', u)$ is invertible for $u' \preceq u \prec c$ because it is an isomorphism by assumption.

Now, we define α on L separately for u smaller than $\underline{c}$, for u between $\underline{c}$ and $\bar{C}$, and for u greater than $\bar{C}$: for $u = s\mathbf{m} + b$ and $v = (s + \eta)\mathbf{m}' + b'$,

$$\alpha(u) = \begin{cases} \psi_{\mathbf{M}}(\pi_{L'}^\downarrow(\underline{c}), v) \circ \varphi_{\mathbf{M}}(\pi_{L'}^\downarrow(\underline{c}), \underline{c})^{-1} \circ \varphi_{\mathbf{M}}(u, \underline{c}) & \text{if } u \prec \underline{c}, \\ \varphi_{\mathbf{M}}(\pi_{L'}^\uparrow(u), v) \circ \varphi_{\mathbf{M}}(u, \pi_{L'}^\uparrow(u)) & \text{if } \underline{c} \preceq u \preceq \bar{C}, \\ \varphi_{\mathbf{M}}(\pi_{L'}^\uparrow(\bar{C}), v) \circ \varphi_{\mathbf{M}}(\bar{C}, \pi_{L'}^\uparrow(\bar{C})) \circ \varphi_{\mathbf{M}}(\bar{C}, u)^{-1} & \text{if } \bar{C} \prec u. \end{cases}$$

The idea of the construction of α is illustrated in Fig. 2. The fact that transition maps of persistence modules only shift forward explains the asymmetry between the cases when $u \prec \underline{c}$ and $\bar{C} \prec u$ in the definition of α.

Let us see that α is well defined. In the case when $u \prec \underline{c}$, α is well defined because $\underline{c} \preceq c$ and for $u \prec c$ the considered maps are invertible by assumption. In the case when $\underline{c} \preceq u \preceq \bar{C}$, we need to show that $\pi_{L'}(u) \preceq v$. Let $u = s\mathbf{m} + b$ and $\bar{u} = \pi_{L'}^\uparrow(u) = t\mathbf{m}' + b'$. It is sufficient to show that $t \leq s + \eta$ because $\hat{m}' > 0$. By definition of $\pi_{L'}^\uparrow$, there exists ι such that $u_\iota = \bar{u}_\iota$. Thus,

$$t - s = \frac{u_\iota - b'_\iota}{m'_\iota} - \frac{u_\iota - b_\iota}{m_\iota} = \frac{u_\iota(m_\iota - m'_\iota) + b_\iota(m'_\iota - m_\iota) + m_\iota(b_\iota - b'_\iota)}{m_\iota \cdot m'_\iota}$$

$$\leq \frac{(|u_\iota| + \|b\|_\infty) \cdot \|\mathbf{m} - \mathbf{m}'\|_\infty + \|m_\iota\|_\infty \cdot \|b - b'\|_\infty}{\hat{m} \cdot \hat{m}'}.$$

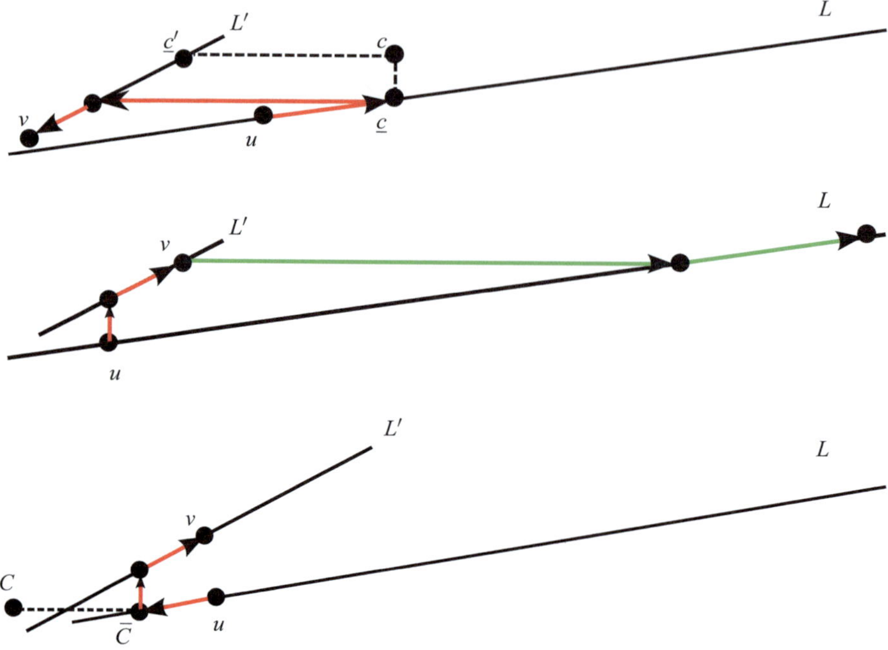

Fig. 2 The map α defined in Lemma 2 is given by composition of the red maps. Top: the case $u \prec \underline{c}$. Bottom: the case $\bar{C} \prec u$. Center: the case $\underline{c} \preceq u \preceq \bar{C}$. The composition of the green maps gives β

The assumption $\underline{c} \preceq u \preceq \bar{C}$ implies $|u_\iota| \leq \max\{\|c\|_\infty, \|C\|_\infty\}$. Thus,

$$t \leq s + \frac{(\max\{\|c\|_\infty, \|C\|_\infty\} + \|b\|_\infty) \cdot \|\mathbf{m} - \mathbf{m}'\|_\infty + \|\mathbf{m}_\iota\|_\infty \cdot \|b - b'\|_\infty}{\hat{m} \cdot \hat{m}'} = s + \eta.$$

(1)

Finally, in the case when $\bar{C} \prec u$, we need to show that $\varphi_{\mathbf{M}}(\bar{C}, u)$ is invertible and $\pi_{L'}^{\uparrow}(\bar{C}) \preceq v$. If $\bar{C} \prec u$, then $\varphi_{\mathbf{M}}(\bar{C}, u)$ is invertible because $C \preceq \bar{C}$ and, by assumption, $\varphi_{\mathbf{M}}(u, u')$ is an isomorphism for every $u, u' \in \mathbb{R}^n$ with $C \preceq u \preceq u'$. The fact that $\pi_{L'}^{\uparrow}(\bar{C}) \preceq v$ follows from the observation that if $\bar{C} = \bar{s}\mathbf{m} + b$ and $\pi_{L'}^{\uparrow}(\bar{C}) = \bar{t}\mathbf{m} + b$, then $\bar{t} \leq \bar{s} + \eta$ as a particular case of (1). Hence, because $\bar{s} < s, \bar{t} \leq s + \eta$, implying that $\pi_{L'}^{\uparrow}(\bar{C}) \preceq v$. Therefore, α is well defined. We can analogously define β.

Finally, it certainly holds that α and β are morphisms of persistence modules, and $\beta(s + \eta) \circ \alpha(s) = \varphi_{\mathbf{M}_L}(s, s + 2\eta)$ and $\alpha(s + \eta) \circ \beta(s) = \varphi_{\mathbf{M}_{L'}}(s, s + 2\eta)$ because α and β have been defined using $\varphi_{\mathbf{M}}$ itself which is a morphism of persistence modules. $\qquad\square$

In order to measure how close L and L' are, it is convenient to confine ourselves to *normalized parametrizations* of L and L': we take $\mathbf{m}, b$ and $\mathbf{m}', b'$ such that

$\|\mathbf{m}\|_\infty = \|\mathbf{m}'\|_\infty = 1$, $\hat{m} = \min_i m_i > 0$, $\hat{m}' = \min_i m_i' > 0$, and $\sum_i b_i = \sum_i b_i' = 0$. This way, $\max\{\|\mathbf{m} - \mathbf{m}'\|_\infty, \|b - b'\|_\infty\}$ quantifies how close L and L' are.

Theorem 2 *Let* $\mathbf{M}$ *be a pointwise finite dimensional persistence module. Assume that there exist* $c \prec C \in \mathbb{R}^n$ *such that* $\varphi_{\mathbf{M}}(u, u')$ *is an isomorphism for every* $u, u' \in \mathbb{R}^n$ *with* $C \preceq u \preceq u'$ *or* $u \preceq u' \preceq c$. *Then, for* L *and* L' *endowed with normalized parametrizations* $u = s\mathbf{m} + b$ *and* $u = s\mathbf{m}' + b'$, *respectively, if* L' *is sufficiently close to* L *so that* $\max\{\|\mathbf{m} - \mathbf{m}'\|_\infty, \|b - b'\|_\infty\} \leq \varepsilon$, *it holds that*

$$d_B(B(\mathbf{M}_L), B(\mathbf{N}_L)) \leq \varepsilon \cdot \frac{\max\{\|c\|_\infty, \|C\|_\infty\} + \|b\|_\infty + \varepsilon + 1}{\hat{m} \cdot (\hat{m} - \varepsilon)}.$$

Proof By Lemma 2, $\mathbf{M}_L$ and $\mathbf{M}_{L'}$ are η-interleaved with

$$\eta = \frac{\left(\max\{\|c\|_\infty, \|C\|_\infty\} + \max\{\|b\|_\infty, \|b'\|_\infty\}\right) \cdot \|\mathbf{m} - \mathbf{m}'\|_\infty + \|b - b'\|_\infty}{\hat{m} \cdot \hat{m}'}$$

$$\leq \frac{\left(\max\{\|c\|_\infty, \|C\|_\infty\} + \|b\|_\infty + \varepsilon\right) \cdot \varepsilon + \varepsilon}{\hat{m} \cdot (\hat{m} - \varepsilon)}.$$

Hence, the claim follows from the Algebraic Stability Theorem for Barcodes. $\qquad\square$

5 Conclusions

In this paper, we have shown that the multidimensional matching distance between rank invariants provides a lower bound for the interleaving distance between multidimensional persistence modules. The proof is by restriction of the persistence modules to increasing lines. The result has an interest in itself because it provides a computable estimate of the interleaving distance. Moreover, it also implies a new proof of the rank invariant stability with respect to perturbations of filtrations. However, this proof if slightly less general than previous ones in that it applies only to data that produce pointwise finite dimensional persistence modules, while already existing proofs apply to more general q-tame persistence modules. Finally, again under the pointwise finite dimensional assumption, also the internal stability of rank invariants is proven in terms of interleavings.

The most natural question left open by this work is whether better bounds on the interleaving distance could be obtained by other distances on rank invariants. Another stable distance that discriminates rank invariants at least as much as the multidimensional matching distance is the *coherent matching distance* introduced in [12], making it a candidate for a more precise estimate of the interleaving distance. However, so far the coherent matching distance is not readily computable, leaving the multidimensional matching distance as the main available tool for this task.

References

1. A. Adcock, D. Rubin, G. Carlsson, Classification of hepatic lesions using the matching metric. Comput. Vis. Image Underst. **121**, 36–42 (2014)
2. U. Bauer, M. Lesnick, Induced matchings of barcodes and the algebraic stability of persistence. J. Comput. Geom. **6**(2), 162–191 (2015)
3. S. Biasotti, A. Cerri, P. Frosini, D. Giorgi, C. Landi, Multidimensional size functions for shape comparison. J. Math. Imaging Vision **32**(2), 161–179 (2008)
4. S. Biasotti, A. Cerri, D. Giorgi, Robustness and modularity of 2-dimensional size functions - an experimental study, in *Computer Analysis of Images and Patterns*, ed. by P. Real, D. Diaz-Pernil, H. Molina-Abril, A. Berciano, W. Kropatsch. Lecture Notes in Computer Science, vol. 6854 (Springer, Berlin, 2011), pp. 34–41
5. S. Biasotti, A. Cerri, P. Frosini, D. Giorgi, A new algorithm for computing the 2-dimensional matching distance between size functions. Pattern Recogn. Lett. **32**(14), 1735–1746 (2011)
6. S. Biasotti, A. Cerri, D. Giorgi, M. Spagnuolo, PHOG: photometric and geometric functions for textured shape retrieval. Comput. Graphics Forum **32**(5), 13–22 (2013)
7. G. Carlsson, Topology and data. Bull. Am. Math. Soc. **46**(2), 255–308 (2009)
8. G. Carlsson, A. Zomorodian, The theory of multidimensional persistence. Discret. Comput. Geom. **42**(1), 71–93 (2009)
9. G. Carlsson, G. Singh, A. Zomorodian, Computing multidimensional persistence. J. Comput. Geom. **1**(1), 72–100 (2010)
10. A. Cerri, C. Landi, Hausdorff stability of persistence spaces. Found. Comput. Math. **16**(2), 343–367 (2016)
11. A. Cerri, B. Di Fabio, M. Ferri, P. Frosini, C. Landi, Betti numbers in multidimensional persistent homology are stable functions. Math. Methods Appl. Sci. **36**, 1543–1557 (2013)
12. A. Cerri, M. Ethier, P. Frosini, The coherent matching distance in 2D persistent homology, in *Computational Topology in Image Context. CTIC 2016*, ed. by A. Bac, J.L. Mari. Lecture Notes in Computer Science, vol. 9667 (2016), pp. 216–227
13. M. Lesnick, The theory of the interleaving distance on multidimensional persistence modules. Found. Comput. Math. **15**(3), 613–650 (2015)
14. S. Oudot, *Persistence Theory: From Quiver Representations to Data Analysis*. AMS Mathematical Surveys and Monographs, vol. 209 (American Mathematical Society, Providence, 2015)
15. K. Xia, G.W. Wei, Multidimensional persistence in biomolecular data. J. Comput. Chem. **36**(20), 1502–1520 (2015)

Persistent Homology over Directed Acyclic Graphs

Erin Wolf Chambers and David Letscher

Abstract We define persistent homology groups over any set of spaces which have inclusions defined so that the corresponding directed graph between the spaces is acyclic as well as along any subgraph of this directed graph. This method simultaneously generalizes standard persistent homology, zigzag persistence, and multidimensional persistence to arbitrary directed acyclic graphs (DAGs), and it also allows the study of more general families of topological spaces or point-cloud data. We give an algorithm to compute the persistent homology groups simultaneously for all subgraphs which contain a single source and a single sink in $O(n^4)$ arithmetic operations, where n is the number of vertices in the graph. We then demonstrate as an application of these tools a method to overlay two distinct filtrations of the same underlying space, which allows us to detect the most significant barcodes using considerably fewer points than standard persistence.

1 Introduction

Since its introduction [15], the concept of topological persistence has found numerous applications in diverse areas such as surface reconstruction, sensor networks, bioinformatics, and cosmology. For completeness, we briefly survey some results from persistent homology with an emphasis on tools and techniques used in this paper, although a full coverage is beyond the scope of this paper. See any of the recent books or surveys on topological persistence for full coverage of this broad topic and its applications [13, 14, 17, 25, 28].

At a high level, the various models of persistence all have a collection of spaces and inclusions of one space into another. The aim is to find topological features common to subsets of these spaces. In standard persistent homology [15], the spaces are linearly ordered with one space included in the next. In zigzag persistence[5],

E. W. Chambers · D. Letscher (✉)
Department of Computer Science, Saint Louis University, Saint Louis, MO, USA
e-mail: echambe5@slu.edu; letscher@slu.edu

© The Author(s) and the Association for Women in Mathematics 2018
E. W. Chambers et al. (eds.), *Research in Computational Topology*, Association
for Women in Mathematics Series 13, https://doi.org/10.1007/978-3-319-89593-2_2

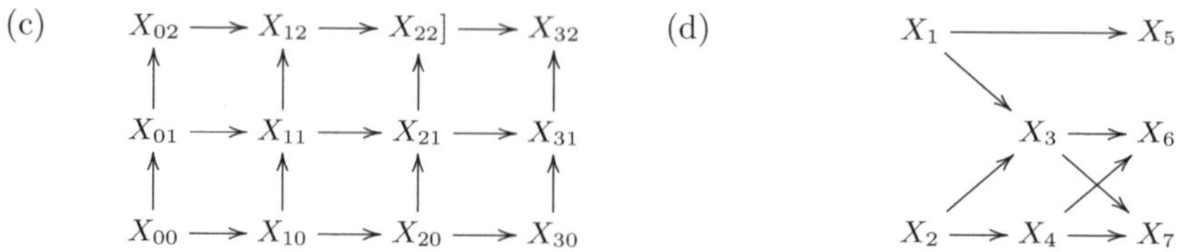

Fig. 1 The underlying graph structure for (**a**) standard persistence, (**b**) zigzag persistence, (**c**) multidimensional persistence, and (**d**) directed acyclic graph (DAG) persistence

the spaces are linearly ordered but the inclusions can occur in either direction. Multidimensional persistence [6] works in multiple dimensions on a grid with inclusion maps parallel to the coordinate axes. In this paper, we present a natural extension of each of these to a directed acyclic graph (or DAG) structure on the spaces and maps, which we call DAG persistence. See Fig. 1 for an example of each of these structures.

Standard persistence considers spaces with maps of the form $X_1 \rightarrow X_2 \rightarrow \ldots \rightarrow X_n$, where each X_i is a topological space, often represented as a simplicial complex. These maps between the spaces induce maps between chain complexes which pass to homology as homomorphisms $H(X_1) \rightarrow H(X_2) \rightarrow \ldots \rightarrow H(X_n)$. Persistent homology identifies homology classes that are "born" at a certain location in the filtration and "die" at a later point. These identified cycles encompass all of the homological information in the filtration and have a module structure [29]. This persistence module has a unique decomposition into a sum of "elementary" modules, which are intervals that share a common homological feature. The endpoints of these intervals are the birth and death times. The set of intervals gives a barcode representation of the persistence module. Persistent homology algorithms have been implemented very efficiently, since the homology groups with coefficient in a finite field form vector spaces, and the inclusion maps induce linear maps between the spaces.

Zigzag persistence considers spaces with maps of the form $X_1 \leftrightarrow X_2 \leftrightarrow \ldots \leftrightarrow X_n$, where the maps can go in either direction. These maps between the spaces induce maps between chain complexes which pass to homology as homomorphisms $H(X_1) \leftrightarrow H(X_2) \leftrightarrow \ldots \leftrightarrow H(X_n)$; this is known as a zigzag module. The zigzag module has the same structure as the persistence module: it has a unique decomposition into a sum of "elementary" modules, which are intervals that share a common homological feature. Like standard persistence, these intervals give a barcode representation of the zigzag module. Recent work in this setting includes an algorithm which examines the order of the necessary matrix multiplications quite carefully and is able to get a running time for a sequence of n simplex deletions or additions which is dominated by the time to multiply two $n \times n$ matrices [22].

In multidimensional persistence, the spaces $\{X_u\}_{u \in \mathbb{Z}^d}$ lie on an integer lattice with inclusion maps for $X_u \to X_v$ where $u, v \in \mathbb{Z}^d$ differ in a single coordinate. Multidimensional persistence modules have a more complicated structure than standard or zigzag, so its interpretation is far more difficult. In particular, no barcode representation exists for multidimensional persistence. One of the primary tools is the rank invariant, $\rho_{X,k}(u, v)$, which measures the rank of homology groups in common among all X_w with $u_i \le w_i \le v_i$.

Our Contribution In this paper, we give a generalization of persistence to spaces where the underlying inclusions form a DAG. This simultaneously generalizes both zigzag and multidimensional persistence, which can be viewed as special cases of these underlying graphs on the maps between the spaces.

To define persistence modules over DAGs, we expand upon the use of quiver modules, first applied to topological persistence in the original zigzag persistence paper [5], to incorporate commutativity conditions that arise from following different paths in a DAG between the same pair of vertices. Note that the authors in [16] examined other ways of incorporating commutativity conditions in specific families of graphs. Given a persistence module, we utilize techniques from category theory, namely limits and co-limits, to define persistence groups for any subgraph of a DAG.

We then give algorithms to compute the persistent homology for DAGs in various settings. In Sect. 4, for graphs with at most n vertices, we give an $O(n^4)$ algorithm for computing the persistent homology groups for all single-source single-sink subgraphs of a DAG simultaneously. In Sect. 5, we describe theoretical algorithms for computing persistence homology for general subgraphs.

Potential applications of this are extensive, including any spaces where inclusions are more general than previous settings. We present two such applications in Sect. 6. The first uses multiple samples of the same space to accurately find significant topological features with far fewer sample points than other methods require. The second application uses DAG persistence to measure the similarity between two spaces.

2 Definition

We recall some relevant definitions and background before presenting our definition of persistent homology over DAGs. For a full presentation of homology groups, see any introductory text in algebraic topology, e.g., [18, 24].

For a simplicial complex X and an Abelian group A, a k-chain is a formal linear combination of the k-simplices of X with coefficients from A; these k-chains form a group which we denote $C_k(X, A)$ (where we will generally omit the A if the group is clear from context). The map $\partial_k : C_k(X) \to C_{k-1}(X)$ is a linear map that calculates the boundary of a chain. The cycle group is defined as $Z_k(X) = \{c \in C_k(X) \mid \partial_k(c) = 0\} = ker(\partial_k)$ and the boundary group is the group $B_k(X) = \{c \in C_k(X)| \exists d \in C_{k+1}(X) \text{ with } \partial_{k+1}(d) = c\} = im(\partial_{k+1})$. The homology group is defined as $H_k(X) = Z_k(X)/B_k(X)$. Note that if A is a field, then $C_k(X)$, $Z_k(X)$, $B_k(X)$, and $H_k(X)$ are all vector spaces.

Given a filtration $X_0 \subset X_1 \subset \cdots \subset X_n$, the persistent homology group $H_k^p(X_j)$ can be defined in multiple ways. Traditionally, it is defined as $Z_k(X_j)/B_k(X_j) \cap Z_k(X_{j+p})$ and can be viewed as a quotient group of $H_k(X_{j+p})$. An equivalent definition is $H_k^p(X_j) = im(i_*)$, where $i_* : H_k(X_j) \rightarrow H_k(X_{j+p})$ is the map induced by the inclusion $i : X_j \rightarrow X_{j+p}$. We note then that $H_k^p(X_j)$ can also be thought of as a subgroup of $H_k(X_{j+p})$.

2.1 Graph Filtrations

In a sense, the persistence group for some interval of a filtration can be thought as the subgroups which are common to all of the homology groups. This motivates the definition of persistence groups over more general sets of inclusion maps; indeed, zigzag persistence and multidimensional persistence are examples of this type of generalization. In this paper, we generalize to consider inclusions over a set of spaces that form a directed graph. The main restriction we will place on this graph is that it must be acyclic and not contain repeated edges, which is a natural constraint for any graph which represents a set of inclusions. Note that we could have equivalently defined these filtrations using a poset, but prefer to use the graph theory terminology. More formally:

Definition 2.1 *For a simple DAG $G = (V, E)$, a* graph filtration *$\mathcal{X}_G$ of a topological space X is a pair $(\{X_v\}_{v \in V}, \{f_e\}_{e \in E})$ such that*

1. *$X_v \subset X$ for all $v \in V$.*
2. *If $e = (v_1, v_2) \in E$, then $f_e : X_{v_1} \rightarrow X_{v_2}$ is a continuous embedding (or inclusion) of X_{v_1} into X_{v_2}.*
3. *The diagram commutes: in other words, suppose we have a path $\gamma = e_1, \ldots, e_l$, we can naturally extend this to a function on the topological spaces $f_\gamma = f_{e_k} \circ \cdots \circ f_{e_1}$. Then, given γ and γ' which are two different directed paths connecting vertices u and w, commuting means that $f_\gamma = f_{\gamma'}$.*

2.2 Persistence Module

Before defining the persistent homology groups for a DAG, we will first generalize the persistence module [29] and zigzag persistence module [5]. We will use an approach similar to that used in the definition of the zigzag persistence module that utilizes quivers and their representations [1, 23]. See [25] for a thorough discussion of the connections between quivers, their representations, and persistent homology.

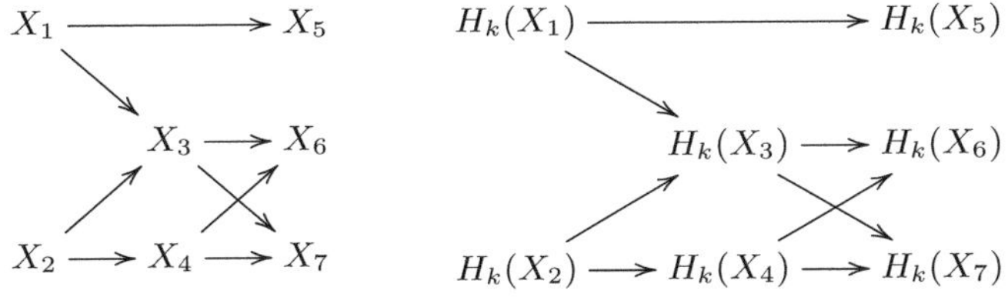

Fig. 2 A graph filtration over a graph G and the corresponding commutative G-module

A *quiver* is a directed graph where loops and multiple edges between the same vertices are allowed. So, every DAG is also a quiver. Given a quiver G, a *representation* of G imparts every vertex of G with a vector space and every edge a linear map between the vector spaces at its endpoints. A representation of a quiver G is also referred to as a G-module.

In order to include the commutativity conditions that are part of a graph filtration, we must use a *quiver with relations*. Formally, a quiver with relations is defined in terms of an ideal of an associative algebra associated to the quiver that is called the path algebra, see [1]. However, a representation of a quiver with relations can be defined without the use of a path algebra. We will focus only on commutativity relations, but more general relations can also be dealt with in a similar manner. This motivates the definition of a commutative G-module, which is a representation of a quiver with (commutativity) relations; see Fig. 2 for an example of such a space.

Definition 2.2 *For a DAG $G = (V, E)$, a* commutative G-module *is the pair* $(\{W_v\}_{v \in V}, \{f_e\}_{e \in E})$ *where for each vertex v, W_v is a vector space and for any edge $e = (v, w)$, $f_e : W_v \to W_w$ is a linear map with the condition that the resulting diagram is commutative.*

This definition provides the framework for discussing the persistence module for a graph that extends the definition for the zigzag persistence module, which will be shown in Sect. 3.2.

Definition 2.3 *For a DAG $G = (V, E)$ and k-dimensional persistence module for a graph filtration $\mathcal{X}_G$, $\mathcal{PH}_k(\mathcal{X}_G)$ is the commutative G-module $(\{W_v\}_{v \in V}, \{f_e\}_{e \in E})$ where*

- $W_v = H_k(X_v)$ *for all $v \in V$.*
- *For every edge $(u, v) \in E$, $f_e : H_k(X_u) \to H_k(X_v)$ is the map induced by the inclusion $X_u \to X_v$.*

The theory for commutative G-modules is very similar to that of zigzag persistence modules. A commutative G-module $\mathcal{V} = (\{V_v\}, \{g_e\})$ is a *submodule* of $\mathcal{W} = (\{W_v\}, \{f_e\})$ if $V_v \subset W_v$ for all v and $f_e|_{V_v} = g_e$. Similarly, given two commutative G-modules $\mathcal{V} = (\{V_v\}, \{f_e\})$ and $\mathcal{W} = (\{W_v\}, \{g_e\})$, we can define their connected sum $\mathcal{V} \oplus \mathcal{W}$ as $(\{V_v \oplus W_v\}, \{f_e \oplus g_e\})$. A commutative G-module is said to be *indecomposable* if it cannot be written as a nontrivial connected sum. The commutative G-modules that we are considering are finite, so $\mathcal{V}$ can be *decomposed*

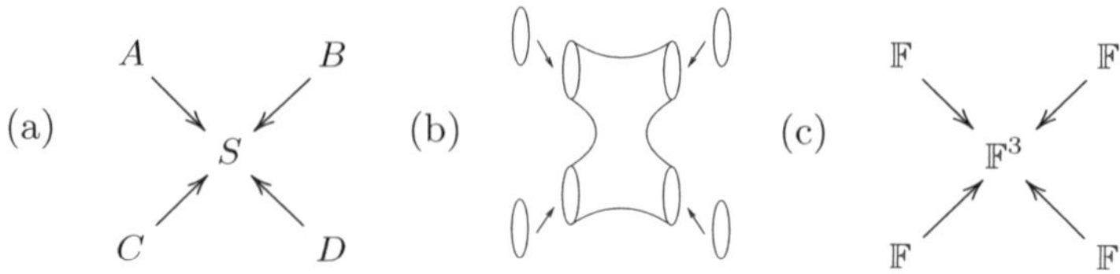

Fig. 3 (**a**) The persistence module for the graph G is not finite type; there are infinitely many non-isomorphic irreducible G-modules. (**b**) A graph filtration $\mathcal{X}_G$. (**c**) The persistence module for $\mathcal{X}_G$ which is irreducible and nonelementary

as $\mathcal{V} = \mathcal{V}_1 \oplus \cdots \oplus \mathcal{V}_n$, where each $\mathcal{V}_i$ is indecomposable. For a connected subgraph G' of G, we will define the commutative G-module $\mathbb{F}_{G'}$ as the module with a copy of $\mathbb{F}$ at each vertex of G' and zero elsewhere; we will put the identity map on each edge of G' and make every other map trivial. We will call this module *elementary*.

Theorem 2.4 (Krull–Remak–Schmidt Theorem for Finite-Dimensional Algebras [19]) *The decomposition of a commutative G-module is unique up to isomorphism and permutation of the summands.*

In the case of standard and zigzag persistence, the relevant indecomposable modules are always elementary. This is implied by Gabriel's theorem [12] which provides an enumeration of indecomposable modules for particular graph types. Figure 3 gives an example of an indecomposable module that is not elementary. The example consists of a sphere with four punctures, and the inclusion of each of the four boundary components. Unfortunately, the existence of such examples tells us that there is no simple "barcode" representation for DAG persistence. In Sect. 2.4, we will generalize barcodes for our context.

2.3 Persistent Homology

We will define persistent homology groups for any connected subgraph of G to include homological features that are present in the homology groups at every vertex of the subgraph. To make this precise, we will define the persistence groups in terms of the limit and co-limit of a commutative G-module; see any text on category theory for a complete discussion of limit and co-limits, for example [21].

In category theory, limits and co-limits are defined for diagrams, a very general concept. In particular, commutative diagrams with every vertex having an object and every edge a morphism are diagrams in the sense of category theory [21]. In particular, commutative G-modules are diagrams. We will give the definitions of limits and co-limits and related structures in terms of commutative G-modules; however, these definitions are identical to those for arbitrary diagrams in [21].

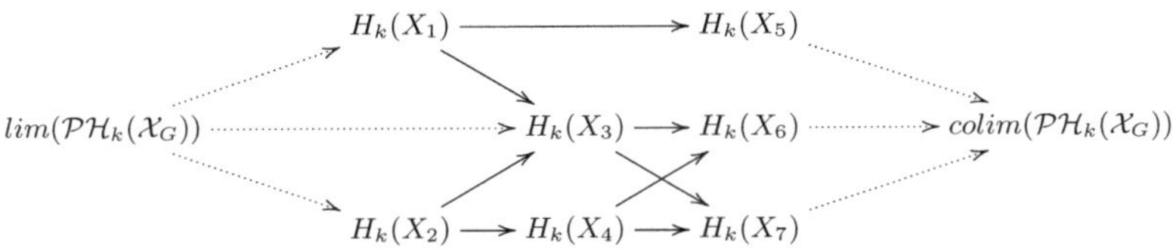

Fig. 4 The limit, $L_{\mathcal{X}_G}$, and co-limit, $C_{\mathcal{X}_G}$, of a diagram of homology groups (For clarity, only maps from the limit to source vertices and maps from sink vertices to the co-limit are shown)

The *cone* of the commutative G-module is a pair, (L, ϕ), with a vector space L and homomorphisms $\phi_v : L \to H_k(X_v)$, such that for any edge (u, v), $f_{uv} \circ \phi_u = \phi_v$. The *limit* of a commutative G-module is a cone (L, ϕ) such that for any other cone (L', ϕ'), there is a unique homomorphism $u : L' \to L$ such that $\phi_e \circ n = \phi'_e$ for every edge e of G. Similarly, a *co-cone* of a commutative G-module is a pair (C, ψ), with a vector space C and homomorphism $\psi_v : H_k(X_v) \to C$ such that for any edge (u, v), $\psi_v \circ f_{uv} = \psi u$; the *co-limit* is a co-cone (C, ψ) of the commutative G-module such that for any other co-cone (C', ψ') there is a unique homomorphism $u : C \to C'$ such that $u \circ \psi_e = \psi'_e$ for every edge e of G. See Fig. 4 for an illustration of the limit and co-limit. When they exist, limits and co-limits are unique up to isomorphism [21]. The existence of the limits and co-limits relies on a few results from category theory, all can be found in [21]. First, G-modules are in the category of vector spaces which are known to be bi-complete, which means that limits and co-limits exist for diagrams from what are known as small categories. Furthermore, commutative diagrams are small categories. Together, these imply the existence of limits and co-limits of commutative G-modules. We will denote the limit and co-limit of M, by $lim(M)$ and $colim(M)$, and let $\mu_M : lim(M) \to colim(M)$ be the induced map between them.

Definition 2.5 *If M is a commutative G-module, then the* persistence *of M, $\mathcal{P}(M)$, is image $\mu_M(lim(M))$.*

The persistence of a module represents all features, or subspaces, common to every vector space M_v in the commutative G-modules. This will be made precise in Lemma 3.3.

Definition 2.6 *Given a graph filtration $\mathcal{X}_G$ and a connected subgraph $G' \subset G$, the G'-persistent homology group, $H_k^{G'}(\mathcal{X}_G)$, is $\mathcal{P}(\mathcal{PH}_k(\mathcal{X}_G)|_{G'})$, where $\mathcal{PH}_k(\mathcal{X}_G)|_{G'}$ is the module $\mathcal{PH}_k(\mathcal{X}_G)$ restricted to the subgraph G'.*

In Sect. 3.2, we connect this definition of persistence to known models and prove that it simultaneously generalizes standard persistence, zigzag persistence, and multidimensional persistence.

2.4 Barcodes and Carrier Subgraphs

We wish to generalize the notion of a persistence barcode to the DAG setting. Every submodule of $\mathcal{PH}(\mathcal{X}_G)$ has a subgraph G' where the module is nontrivial. We will call this the *carrier subgraph* of that module and say that the module is *carried* by that subgraph. In standard and zigzag persistence, all irreducible modules are elementary, and a barcode is precisely a carrier subgraph of this module. In our more general setting, we cannot assume that irreducible modules are elementary. In fact, these irreducible submodules can be very complicated. For example, in Fig. 3 the carrier subgraph would be the entire graph. Even though carrier subgraphs for general DAGs are not a complete invariant, that is not all information is encoded in these representations, they can still provide insight into the structure of the persistence module.

2.5 An Example

In Fig. 5, we see an example of a set of spaces with inclusions that form a DAG. At the top level is a genus two surface; the directed arrows indicate the inclusion maps in our DAG, down to our two source vertices in the graph which include one space with two disjoint annuli and one space that is a disk with three additional boundaries. The graph forms a poset that demonstrates the nontrivial intersections and unions of the three surfaces. In Fig. 5b, we see the persistence module for the entire space, and in Fig. 5c, d we see the indecomposable submodules and their carrier subgraphs. From these indecomposables, it is possible to read off the persistence for any subgraph G' by counting how many of the elementary modules have G' as a subgraph.

3 Properties of Persistence Module

Before examining algorithms to calculate persistence and persistence modules over DAGs, we will examine various properties of commutative G-modules and their relationship with the persistence module and zigzag persistence module.

3.1 Fundamental Properties

There are several properties of commutative G-modules and their persistence that are useful in seeing their relationships with existing models of persistence and how they can be computed. The first shows how the persistence of a commutative G-module, M, carries information common to every vertex group M_v.

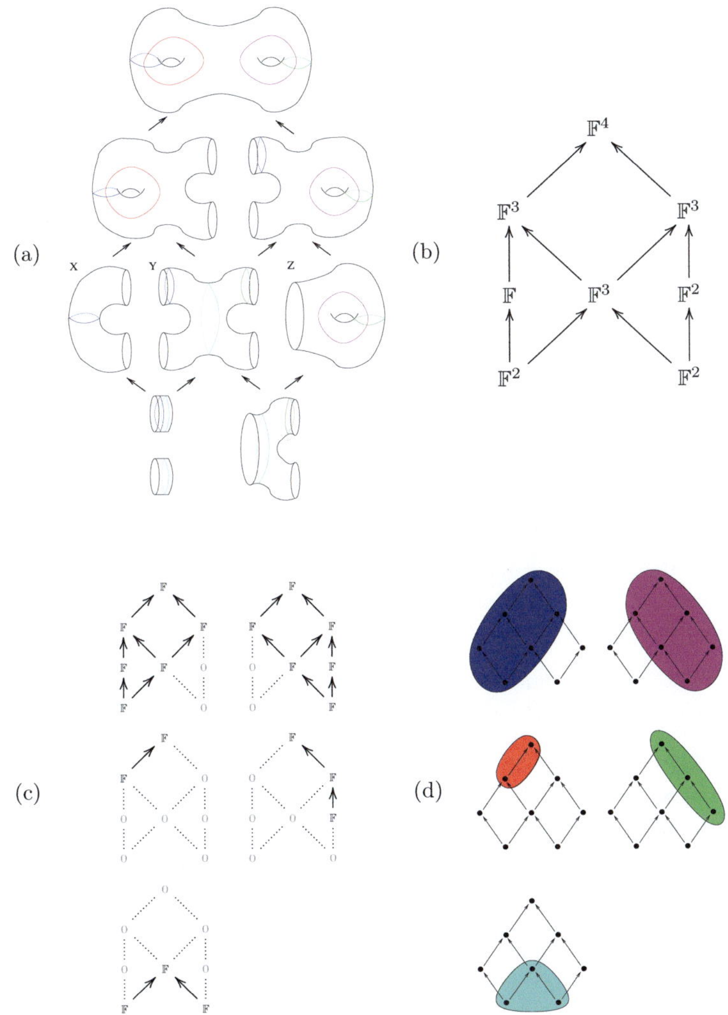

Fig. 5 (**a**) A genus two surface that is the union of three subsurfaces (X, Y, and Z), with maps between subspaces forming a directed acyclic graph (DAG) and generators indicated by colored cycles, (**b**) its persistence module $\mathcal{PH}_1(\mathcal{X}_G)$, (**c**) the five indecomposable summands of the persistence module (indicated in bold), and (**d**) the five carrier subgraphs (marked by ovals and shaded to match colors of generators in part a) for this example

Lemma 3.1 *If M is a commutative G-module, then*

$$M \cong \mathcal{P}(M)_G \oplus N$$

where $\mathcal{P}(M)_G$ is the module with a copy of $\mathcal{P}(M)$ at each vertex of G and $\mathcal{P}(N) = 0$.

Proof Consider the linear map $\mu_m : lim(M) \to colim(M)$. This gives two decompositions

$$lim(M) \cong im(\mu_M) \oplus ker(\mu_m) = \mathcal{P}(M) \oplus ker(\mu_m)$$

$$colim(M) \cong im(\mu_M) \oplus cok(\mu_m) = \mathcal{P}(M) \oplus cok(\mu_m)$$

where $cok(\mu_M) = N/im(\mu_m)$. So, there is a copy of $\mathcal{P}(M) \subset lim(M)$ that maps isomorphically to $\mathcal{P}(M) \subset colim(M)$. Consider the limit of M, $\mathcal{P}(M)$, and the maps $\phi_v : \mathcal{P}(M) \to M_v$. This decomposes $M_v \cong P \oplus cok(\phi_v)$. Notice that the maps $M_u \to M_v$ preserve this decomposition. This established that $M \cong \mathcal{P}(M)_G \oplus N$, where $N_v \cong cok(\phi_v)$.

To show that $\mathcal{P}(N) = 0$, note that $\mathcal{P}(M_1 \oplus M_2) \cong \mathcal{P}(M_1) \oplus \mathcal{P}(M_2)$, since limits and co-limits preserve direction sums [21]. If $\mathcal{P}(N) \neq 0$, then

$$\mathcal{P}(M) \cong \mathcal{P}(\mathcal{P}(M)|_G) \oplus \mathcal{P}(N) \cong \mathcal{P}(M) \oplus \mathcal{P}(N) \not\cong \mathcal{P}(M)$$

This contradiction shows that the persistence of N is trivial.   □

Note that the previous implies that for every vertex v of G, there is an injection of $\mathcal{P}(M)$ into M_v. This lemma yields a characterization of when indecomposable submodules are elementary, precisely when their persistence is nontrivial.

Corollary 3.2 *Assume that M is an indecomposable commutative G-module, then $\mathcal{P}(M) \neq 0$ if and only if $M \cong \mathbb{F}_G$.*

Proof Using the decomposition $M \cong \mathcal{P}(M)_G \oplus N$ in the previous lemma, we observe that since M is irreducible one of the terms must be trivial. In one case $M = \mathcal{P}(M)_G = \mathbb{F}_G$ and in the other $\mathcal{P}(M) = 0$.   □

The next lemma provides the basis of both computing DAG persistence on a large class of subgraphs and relating this model of persistence to existing theories.

Lemma 3.3 *If G is a single-source single-sink graph with source and sink vertices s and t, respectively, and $M = (\{M_V\}, \{f_e\})$ is a commutative G-module, then $\mathcal{P}(M) \cong im (M_s \to M_t)$.*

Proof First consider the limit of M. Notice that for any cone $(C, \{g_v\})$ of M, all of the maps $g_v : C \to M_v$ are completely determined by g_s; specifically $g_v = f_{\gamma_{sv}} \circ g_s$ where γ_{sv} is any path in G from s to v. We claim that $(M_s, \{f_{\gamma_{sv}}\})$ is the limit of M.

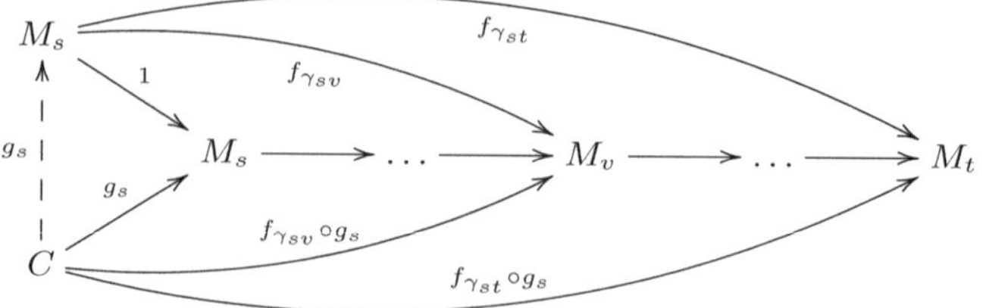

In the above diagram, it is clear that the cone from C factors through the cone M_s.

A similar argument shows that the co-limit of M is $(M_t, \{f_{\gamma_{vt}}\})$. So, $\mathcal{P}(M)$, the image of the limit in the co-limit, is identical to the image $M_s \to M_t$. $\qquad\square$

3.2 Relationship with Other Models of Persistence

We saw in Sect. 2 the definition for standard persistence. Multidimensional persistence can also be defined in terms of the image of a map. Consider a multifiltration, or d-dimensional grid of spaces equipped with a partial ordering on vertices, where $u = (u_1, \ldots, u_d) \leq v = (v_1, \ldots, v_d)$ if and only if $u_i \leq v_i$ for all i.

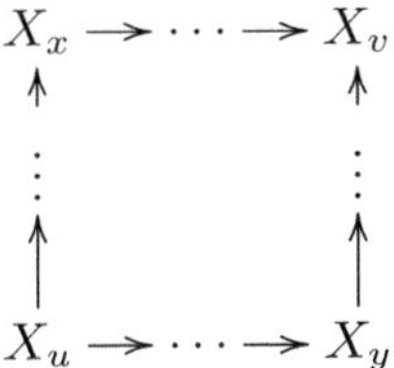

The *rank invariant*, $\rho_{X,k}(u, v)$, is defined as the dimension of the image of the induced map $H_k(X_u) \to H_k(X_v)$ [5]. Since the diagram commutes, this map can be found following any path from u to v in the graph.

Zigzag persistence is defined in terms of the zigzag module. The definition of a commutative G-module and a τ-module for zigzag persistence are identical for a zigzag graph. The following proposition specifies how DAG persistence generalizes these three notions of persistence.

Proposition 3.4 *Suppose $\mathcal{X}_G$ is a graph filtration of X. Then:*

1. *(Standard persistence) If G is the graph corresponding to the filtration $X_0 \to X_1 \to \cdots \to X_n$ and $I_{i,p}$ is the subgraph consisting of vertices $\{X_i, \ldots, X_{i+p}\}$, then $H_k^{I_{i,p}}(\mathcal{X}_G) \cong H_k^p(X_i)$. Furthermore, $\mathcal{PH}_k(\mathcal{X}_G)$ coincides with the persistence module.*
2. *(Zigzag persistence) If G is the graph for zigzag persistence $\mathbb{X} = X_0 \leftrightarrow X_1 \leftrightarrow \cdots \leftrightarrow X_n$ where each arrow could go in either direction, then $\mathcal{PH}_k(\mathcal{X}_G)$ coincides with the zigzag persistence module.*
3. *(Multidimensional persistence) Let $\mathcal{X} = \{X_v\}_{v \in \{0,\ldots,m\}^d}$ be a multifiltration with underlying graph G. If $G_{u,v}$ is the subgraph with vertices $\{w \in G \mid u \leq w \leq v\}$, then the rank invariant $\rho_{X,k}(u, v) = \dim H_k^{G_{u,v}}(\mathcal{X}_G)$.*

Proof In the case of zigzag persistence, the definition of the zigzag persistence module is identical to the definition of the DAG persistence module; in particular for the case of a zigzag graph G the definition of a commutative G-module coincides exactly with the definition of a τ-module [5]. Note that the standard persistence module is a special case of the zigzag persistence module so this also shows that the persistent module coincides with the DAG persistence module when the graph is a single path.

In the case of standard persistence, Lemma 3.3 implies that $H_k^{l_i, p}(\mathcal{X}_G) = im(H_k(X_i) \to H_k(X_{i+p}))$, which is one of the definitions of the (standard) persistent homology group. So, the two ways of calculating persistent homology groups coincide for graphs that are paths.

In the case of multidimensional persistence, $\rho_{X,k}(u, v)$ is defined as the rank of the image $H_k(X_u) \to H_k(X_v)$. Lemma 3.3 can also be applied in this situation to show that $H_k^{G_{u,v}} = im(H_k(X_u) \to H_k(X_v))$. So, the rank invariant is the same as the rank of the (DAG) persistent homology group. $\qquad\square$

4 Single-Source Single-Sink Subgraphs

In this section, we consider $\mathcal{X}_G$ where G is a DAG with a single-source vertex s. Also, each vertex of the graph will represent a subset of a particular fixed cell complex, although this final complex may or may not actually appear as a complex attached to a vertex in G. To limit the number of cells which f_e can introduce (where $e = (u, v) \in G$), we assume that each inclusion f_e adds a single cell to the underlying space. We will also assume at each source vertex s that $X_s = \emptyset$. These assumptions are standard in most persistence algorithms [22, 29] and quite natural given that we can decompose any inclusion map into a series of inclusions of one simplex at a time.

If we consider a subgraph $G' \subset G$ with source s and sink t, Lemma 3.3 implies that we only need to find the image of $H_k(X_s) \to H_k(X_t)$. This image can be found by following any path from s to t and applying the standard persistence algorithm. This can be repeated for every pair of vertices of G. Therefore, a straightforward application of the cubic time standard persistence algorithm would yield an $O(n^2 l^3)$ algorithm, where n is the number of vertices of the graph and l is the length of the longest directed path from source to sink. Using tools from recent work to compute zigzag persistence in matrix multiply time [22] would give a running time of $O(n^2(M(l)) + l^2 \log^2 l))$, where l is the length of the longest path between any source and sink and $M(l)$ is the time to multiply two $l \times l$ matrices.

Here, we will adapt the standard persistence algorithm [15], shown in Algorithm 1, to calculate the persistent homology for all single-source single-sink subgraphs. First, we recall the standard persistence algorithm, which starts with the $(k + 1)$-dimensional boundary map stored in a matrix. This means that there is a column for each $(k + 1)$-cell and a row for each k-cell. Each column stores the boundary of its $(k+1)$-cell, with a 1 or -1 in each entry, where the sign depends on the positive or negative orientation of the k-cell as a face in the larger $(k + 1)$-cell. There is also an additional array that stores the index of the first nonzero entry in each column. (In the original paper, it was assumed that the field was $\mathbb{Z}_2$, but the same algorithm works for any finite field.) For completeness, below is pseudocode for the standard algorithm applied to a boundary matrix B. Note that the entries for the matrix $\{b_{cr}\}$ are stored in column major order and that the i-th column of B will be denoted by B_i.

Algorithm 1 Standard persistence

---

procedure STANDARDPERSISTENCE(B)
    for $c = 1..\#$ columns of B **do**
        for $c' = 1..c - 1$ **do**
            Let r be the first row where $b_{c'r}$ is non-zero
            if $b_{cr} \neq 0$ **then**
                Replace B_c with $B_c - b_{cr}(b_{c'r})^{-1} B_{c'}$
    return B

At its completion, the algorithm terminates with a matrix where the index of each column is the death time of a cycle that has birth time at the index of its first nonzero row. And a basis for $H_k^p(X_i)$ can be extracted from B extracting the columns $1, \ldots, i + p$ whose first nonzero entry has index at least i [15].

Our algorithm to calculate persistence on every single-source single-sink subgraph of G relies on two observations. The first is that the standard persistence algorithm can be modified to run on a tree without increasing its asymptotic runtime. The second is that every DAG can be covered in a linear number of trees so that for any vertices u and v that have a directed path between them in G also have a path between them in at least one of the trees. So, we can call the tree-based persistence algorithm n times and recover all of the persistent homology groups for single-source single-sink subgraphs of G.

To calculate persistence on a tree, T, we make the same assumptions as before, at the root r, $X_r = \emptyset$ and the transition from each edge adds a single cell. Let O be an ordered list of the vertices of T in any topological ordering and let $<_T$ be the partial order for the elements of the tree. The boundary matrix B will be an $|O| \times |O|$ matrix with both rows and columns indexed by O. The entries of the column B_v will be the boundary of the cell σ that is added in the inclusion $X_u \rightarrow X_v$ where (u, v) is an edge of T. Notice that if T was just a path, the matrix B would be the same as the boundary matrix used in the standard persistence algorithm. Below is the algorithm.

Algorithm 2 Persistence calculation on a tree

---

procedure TREEPERSISTENCE($B,O,<_T$)
    for $c \in O$ **do**
        for $c' \in O$ **do**
            if $c <_T c'$ **then**
                Let $r \in O$ be the first row with $b_{c'r}$ non-zero
                if $b_{cr} \neq 0$ **then**
                    Replace B_c with $B_c - b_{cr}(b_{c'r})^{-1} B_{c'}$
    return B, R

If $\gamma \subset T$ is a path from u to v, then a basis for $H_k^\gamma(\mathcal{X}_G)$ can be found by extracting the column from B with column index at most v in the partial ordering $<_T$ and that have their first nonzero entry at least u in the partial ordering.

Lemma 4.1 *If $\{\gamma_v\}$ are paths from the root r to the vertex v in T, then Algorithm 2 correctly finds a basis $H_k^{\gamma_v}(\mathcal{X}_G)$ for all k and all $v \in T$ in $O(n)$ time, where n is the number of vertices of G.*

Proof Notice that entries in column v and row u of B are always 0 if u and v are not connected by a directed path in T. In the calculation of column v of the matrix, only columns u with $u \leq_T v$ can affect the values in the column. If the calculations involving the other columns are ignored, then the calculation is identical to the one done in the standard persistence algorithm (with extraneous zeros) and returns the same result.

The algorithm clearly runs in $O(n^3)$ time since each loop proceeds through at most n elements and the body of the inner loop takes linear time to scan for the first nonzero elements and update the row. $\qquad\square$

To calculate persistence for all single-source single-sink subgraphs, we construct n trees, one for each vertex v in the graph. The tree starts with a path from a source vertex to v combined with a tree that spans all vertices u where there is a path u to v. Since the persistent homology for these graphs does not depend on the path taken, running the tree algorithm on this tree will always calculate persistence for all single-sink single-source subgraphs with source v.

Algorithm 3 Persistent homology for all single-source single-sink subgraphs

---

procedure ALLSINGLESOURNCESINGLESINK($\mathcal{X}_G, V, E, <_G$)
    for $v \in V$ **do**
        Let γ be a path from any source to v
        $U = \{u \in V \mid v \leq_G u\} \cup \gamma$
        Let $T \subset G$ be a tree rooted at v with vertex set U
        Let O be a topological ordering of the vertices of T
        Construct a square matrix B of all zeros indexed by the elements of O
        for $v \in O$ **do**
            Let (u, v) be an edge of T
            Let σ be the simplex of $X_v - X_u$
            Set the column indexed by v to store the entries of $\partial\sigma$
        $B_v = $ TREEPERSISTENCE($B, O, <_T$)
    return $\{B_v\}$

Note: The tree T is a spanning tree for the subgraph of G with vertex set U, the partial ordering and partial order can be constructed in $O(n \log n)$ time at the same time a topological ordering can be constructed using standard graph algorithms, see [11] for details.

Theorem 4.2 *If G is a DAG that has n vertices, then Algorithm 3 calculates all of the persistent homology groups with coefficients from a finite field for all single-source single-sink subgraphs of G can be calculated in $O(n^4)$ time and $O(n^3)$ space.*

Proof Consider a single-source single-sink subgraph G' with source u and sink v. Lemma 3.3 implies that $H_k^{G'}(\mathcal{X}_G) \cong im(H_k(X_u) \to H_k(X_v))$, which we know that Algorithm 2 correctly calculates and stores in B_u.

Algorithm 3 loops through each vertex once. The body of the loop takes $O(n \log n)$ time to construct the tree and associate structures and $O(n^3)$ time for the call to Algorithm 2. This gives a total runtime of $O(n^4)$. Notice that each call to tree persistence uses $O(n^2)$ space so the total space usage is $O(n^3)$. $\qquad\square$

We note that if we remove the assumption that coefficients are from a finite field, our theorem still accurately counts the number of arithmetic operations, although running times for each operation might take longer. In 3 dimensions, however, no information is lost when persistent homology is taken with respect to a finite field [20], so this is not an overly restrictive assumption. We also note that this algorithm can be adapted to multidimensional persistence to give an improvement over the known polynomial time algorithm [7]:

Proposition 4.3 *All of the rank invariants for a d-dimension lattice with n nodes can be calculated in $O(n^{4-1/d})$ time.*

Proof The improvement in the run-time for lattices is that we do not need to run the tree-based algorithm for every vertex of the graph. Instead, we can choose trees for each vertex of the form $v = (0, v_2, \ldots, v_d)$. These trees will follow a path from the origin to v and then go through every vertex $u = (u_1, u_2, \ldots, u_d)$ with $v_i < u_i$ for $i \geq 2$. In addition to the edges in the path to v, there will be an edge from $u' \to u$ whenever u' is the largest vertex in the lexicographic order with $v \leq_G u' \leq_G u$.

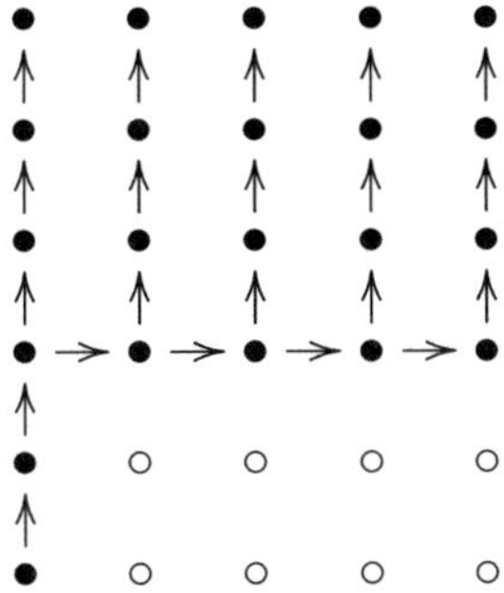

Every pair of vertices that can be connected in the lattice by a directed path can also be connected in one of these trees. There are $n^{\frac{d-1}{d}} = n^{1-1/d}$ such trees, yielding the improved running time. $\qquad\square$

5 Indecomposable Submodules and Persistence over General Subgraphs

For a general subgraph $G' \subset G$, calculating $\mathcal{PH}_k^{G'}(\mathcal{X}_G)$ is much more complicated. However, we can utilize algorithms from computational algebra to solve and calculate DAG persistence for arbitrary subgraphs. To do so, we first need to address decomposing the DAG persistence module into indecomposables.

Algorithm to Decompose the Persistence Module Viewed in other contexts, a DAG persistence module is a representation of a quiver with (commutative) relations or, equivalently, a representation of an associative algebra. Our representations are finite-dimensional and have coefficients in a finite field. Under these conditions, there are polynomial time algorithms to decompose the module into indecompossibles [3, 10]. These, and similar, algorithms have been implemented in the computer algebra systems GAP [27] and MAGMA [4]. For examples that are not too large, these algorithms work well in practice. For larger examples that we attempted, GAP failed to return an answer.

Algorithm to Calculate the Rank of $\mathcal{H}_k^{G'}(\mathcal{X}_G)$ for General Subgraphs Given a decomposition into indecomposable submodules, it is relatively straightforward to calculate the persistence groups for arbitrary subgraphs.

Let P be the restriction of $\mathcal{PH}_k(\mathcal{X}_G)$ to the subgraph G'.
Decompose P into irreducibles $M_1, \ldots, M_n$.
Initialize $r = 0$
for $i = 1..n$ **do**
    if M_i is elementary **then**
        Increment r

Given a representation of a submodule, we can check if it is elementary since at every vertex of the DAG the representation is either rank 0 or 1. Note that Lemma 3.2 implies that each of the M_i are either elementary and contribute exactly one to the rack of the persistence group or have $H_k^{G'}(M_i) = 0$ and contribute nothing to the persistence group.

6 Applications

We finally will demonstrate two proof of concept applications of our definitions, both of which use our single-source single-sink algorithm from Sect. 4. Both examples utilize the same dataset: a piecewise analytic surface of genus two. Point samples were sampled uniformly from the surface.

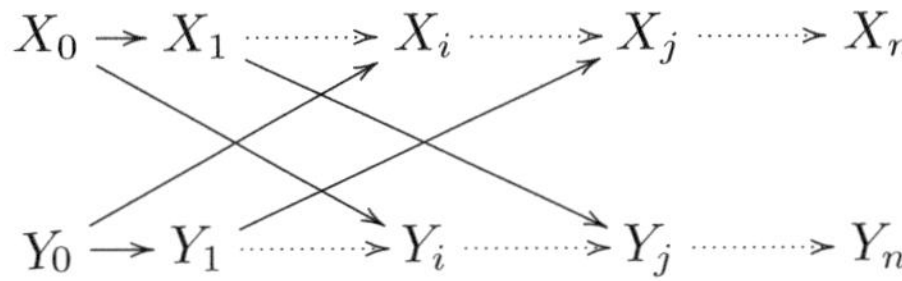

Fig. 6 Filtration of union of balls corresponding to each of the two subsamples where all arrows indicate inclusion maps

6.1 Estimating Persistence Using Multiple Subsamples

The first application is estimating persistence for a point sample using a pair of much smaller subsamples. Let $X_0 \to X_1 \to \cdots \to X_n$ and $Y_0 \to Y_1 \to \cdots \to Y_n$ be filtrations for the union of balls of various radii for two subsamples of a common point set. Moreover, we assume that each of the X_i is contained in some Y_j and vice versa. This yields a directed graph of the form shown in Fig. 6. This is the picture of an ϵ-interleaving [8]. It is known that the persistence diagrams of these two filtrations have distance at most ϵ [9]. Using DAG persistence, we try to exploit additional information encoded in the graph to distinguish the significant cycles from noise.

If we could calculate the carrier graphs for each irreducible submodule for such a dataset, then we could build a "persistence diagram." The birth times would be the average index of the first X_i and Y_i that appear in the carrier graph and the death time would be the average of the largest index. In theory, we could use the algorithms in Sect. 5, but they were too large for GAP to process. Instead, we used an alternative heuristic process:

1. Fix a small constant $s > 0$.
2. Build filtrations $\{X_i\}$ and $\{Y_i\}$ where $X_i = \{x \mid d(x, X_0) \le is\}$ and Y_i is defined similarly.
3. Estimate if $X_i \subset Y_j$ and $Y_i \subset X_j$.

   (a) Decompose the region into a rectangular grid with spacing s.
   (b) Each grid point is labeled by its distance to X_0 and Y_0 which were rounded to their closest multiple of s.
   (c) We will assume that $X_{\alpha s} \subset Y_{\beta s}$ if for every grid points with distance at most αs to X_a has distance to Y_0 at most βs.

4. Calculate the rank of the persistent homology using the single-source single-sink algorithm of Sect. 4 and record their bases.
5. Consider the set of bases and merge any two graphs that have one of the same bases in common.

This heuristic algorithm attempts to construct the carrier subgraphs of the irreducible submodules. This process is not guaranteed to identify the carrier subgraphs of the irreducible submodules. However, in our particular example, we were able to validate the results and show that it correctly identified the carrier subgraphs. We find it interesting that the heuristic was successful in finding the irreducible submodules and wonder how likely this is to occur in practice.

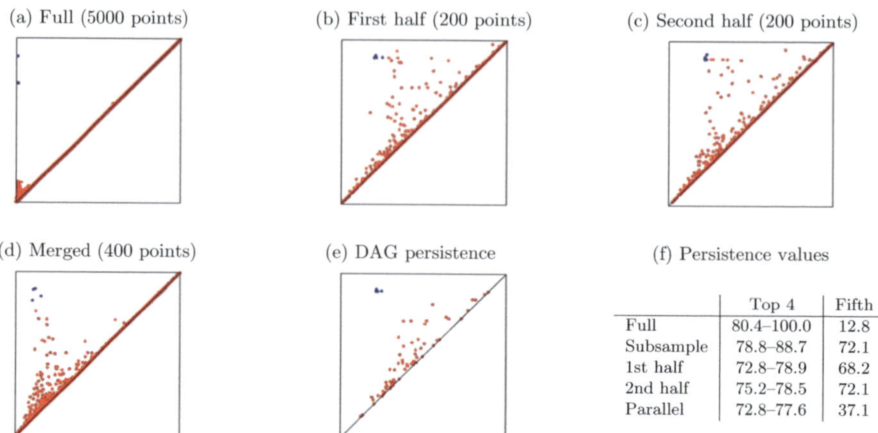

Fig. 7 Persistent homology for a point sample of a double torus. (**a**) Persistence diagram for the original 5000 point sample, (**b**) and (**c**) persistence diagrams for two different 200 point subsamples, (**d**) persistence diagram for the union of the two small samples, (**e**) DAG persistence diagram, and (**f**) the distance from the diagonal to the five most significant features. In all of the persistence diagrams, the four most significant cycles are colored blue to highlight them

In Fig. 7, we show the result of this process. Each subfigure is a persistence diagram, where each cycle is considered as a pair of birth and death times. The persistence of each cycle is the difference in these times which is equal to the distance to the diagonal. The space considered is a genus two surface sampled with 5000 points, at which level the persistent features, which are the 4 generators of homology, are clearly seen as significant in Fig. 7a. Note that these four points blur together in pairs in the figure and are difficult to distinguish from each other. In the remaining pictures, we calculate persistent homology for a simple DAG that consists of two different 200 point subsamples (Fig. 7b, c), and their union into a larger 400 point sample (Fig. 7d). We note that individually, each sample's persistent homology is quite noisy and does not distinguish the four generators at all from the "noise." However, the persistent homology for the DAG (Fig. 7e) clearly separates the 4 main generators from the noise, at a far lower level of sampling than is possible with standard persistent homology. The table (Fig. 7f) shows the persistence values for the four generating cycles for the surface and the next largest cycle. For simplicity of interpretation, the input data was scaled so that the most significant cycle in the 5000 point sample has lifespan equal to 100.

Intuitively, short lived cycles in one of the filtrations do not align well with cycles in the other filtration. So, these cycles are paired with very short cycles in the other filtration. The corresponding points on the persistence diagram are moved toward the diagonal. This deemphasizes the cycles that are the result of sampling noise.

Additional possible applications of this are numerous. For example, we could use DAGs of various witness complexes and seek improved results using a DAG over a small set of witness complexes; recent work using zigzag persistence seems relevant in this setting [26]. Also, the example in Sect. 2.5 demonstrates how homology classes from pieces of a space can be "aligned" similar to the bootstrapping method of [26].

It should be noted that this is not a faster way to calculate the persistence of the larger sample. This experiment was performed to test if the filtrations for two small subsamples could be combined to more accurately represent the topology of the shape than the union of the subsamples. In this case, we were able to answer in the affirmative, but this would not work on all such examples and would be computationally infeasible in practice.

6.2 *Shape Comparison*

Given filtrations of two overlapping shapes, a natural question is to measure how similar they are. DAG persistence can provide a method for comparison. Consider the graph in Fig. 8; it includes filtrations for two shapes X and Y as well as their intersection and union. If X and Y are very similar and well aligned, then this should be detected in the carrier subgraphs.

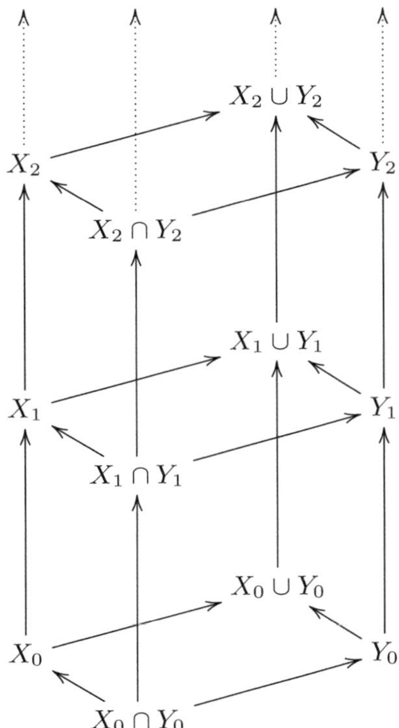

Fig. 8 Underlying graph for the comparison of two filtrations

This comparison is made by building persistence diagrams for each filtration $\{X_i\}$ and $\{Y_i\}$ using standard persistence. A persistence diagram is built for the module over G by first calculating its irreducible modules and their carrier subgraphs. In this example, all of the irreducible subgraphs were elementary making further analysis possible. Each carrier subgraph is converted to a point (i, j) in the persistence diagram if i is the smallest index and j is the largest index of the vertices in the carrier subgraph.

Similar to the previous application, it was not possible to perform this calculation exactly as we have no efficient algorithm to find the irreducible submodules. Instead, we used a very similar heuristic process to the one in the previous example, except we had four filtrations to compare instead of two.

This comparison was performed on two 1000 point subsamples from the dataset in the previous example, see Fig. 7. The bottleneck distance between the persistence diagrams of each of the two samples and the persistence diagram for the comparison graph were both under 5% of the lifespan of the 4 significant topology features of the shapes. This demonstrates that the two shapes being compared have nearly identical topological features except on a very small scale.

An alternate way of comparing the filtration $\{X_i\}$ and $\{Y_i\}$ is to find the interleaving distance between them [8]. This is the minimum ϵ so that for all i, $X_i \subset Y_{i+\epsilon}$ and $Y_i \subset X_{i+\epsilon}$. In general, calculating interleaving distance is very hard to do. The DAG persistence for this graph does not approximate interleaving distance, but it has the potential to be used for some of the same purposes. This is a direction for further study.

7 Future Work

Our algorithm in Sect. 5 is very slow due to the high degree both theoretically and in practice. However, the algorithm to find the decomposition of the persistence module is designed for modules over very general algebras. Significant improvements should be possible that utilize the fact that commutative G-modules have an additional structure. The heuristics used in Sect. 6 used this additional structure, but we have been unable to prove that they will always be successful in correctly decomposing the persistence module.

Recent work has focused on developing parallel frameworks for persistence [2]. A second practical application of our formulation would be using the type of splitting shown is Sect. 2.5 as a basis for a divide and conquer algorithm that allows parallel computation of standard persistence. For example, in Fig. 5, we can calculate the persistence of the entire space by combining the computation for the subspaces X, Y, and Z, which can be done in parallel.

There is the potential to improve our algorithm for all single-source single-sink subgraphs using tools from recent work that computes zigzag homology in matrix multiply time [22]. Consider a directed tree T in the DAG for a single sink s; this tree is composed of paths which look like filtrations used in zigzag persistence, so we can use the recent zigzag algorithm for each path. If this tree has common subpaths from previous calls to the zigzag algorithm, we could potentially speed up our algorithm by using this information. On the other hand, if the tree has no such common subpaths, then intuitively we should be able to balance the fact that the disjoint paths sum to n (so that we either have many very short paths or the tree is a single path). Balancing this recursion based on the analysis in [22], however, gives no better running time than the naive $O(n^2 M(l))$ algorithm we briefly outlined in Sect. 4, since the time to combine the information from a previous call with the new recursive call will take longer than the call itself. A better algorithm that uses this information successfully is an interesting direction to consider.

Finally, the alignment process proposed in Sect. 6.2 should be studied further. It could prove a faster alternative to finding interleavings of filtrations.

Acknowledgements The authors would like to thank Afra Zomorodian for suggesting this problem during a visit as well as for additional comments and suggestions along the way. We would also like to thank Greg Marks and Michael May for helpful conversations during the course of this work. Finally, we would like to thank the anonymous reviewers for a number of suggestions for improvement of this paper.

Research supported in part by the National Science Foundation under Grant No. CCF 1054779 and IIS-1319573.

References

1. M. Barot, Representations of quivers, in *Introduction to the Representation Theory of Algebras* (Springer, Berlin, 2015), pp. 15–31
2. U. Bauer, M. Kerber, J. Reininghaus, Distributed computation of persistent homology. CoRR, abs/1310.0710 (2013)
3. P.A. Brooksbank, E.M. Luks, Testing isomorphism of modules. J. Algebra **320**(11), 4020–4029 (2008)
4. W. Bosma, J. Cannon, C. Playoust, The Magma algebra system. I. The user language. J. Symbolic Comput. **24**(3–4), 235–265 (1997). Computational algebra and number theory (London, 1993)
5. G. Carlsson, V. de Silva, Zigzag persistence. Found. Comput. Math. **10**, 367–405 (2010)
6. G. Carlsson, A. Zomorodian, The theory of multidimensional persistence. Discret. Comput. Geom. **42**(1), 71–93 (2009)
7. G. Carlsson, G. Singh, A. Zomorodian, Computing multidimensional persistence. J. Comput. Geom. **1**(1), 89–119 (2010). Preliminary version appeared in ISAAC 2009
8. F. Chazal, D. Cohen-Steiner, M. Glisse, L.J. Guibas, S.Y. Oudot, Proximity of persistence modules and their diagrams, in *Proceedings of the Twenty-fifth Annual Symposium on Computational Geometry, SCG '09* (ACM, New York, 2009), pp. 237–246
9. F. Chazal, V. De Silva, M. Glisse, S. Oudot, *The Structure and Stability of Persistence Modules* (Springer, Berlin, 2016)
10. A. Chistov, G. Ivanyos, M. Karpinski, Polynomial time algorithms for modules over finite dimensional algebras, in *Proceedings of the 1997 International Symposium on Symbolic and Algebraic Computation* (ACM, New York, 1997), pp. 68–74
11. T.H. Cormen, *Introduction to Algorithms* (MIT Press, Cambridge, 2009)
12. H. Derksen, J. Weyman, Quiver representations. Not. AMS **52**(2), 200–206 (2005)

13. H. Edelsbrunner, J. Harer, Persistent homology—a survey, in *Essays on Discrete and Computational Geometry: Twenty Years Later* ed. by J.E. Goodman, J. Pach, R. Pollack. Contemporary Mathematics, vol. 453 (American Mathematical Society, New York, 2008), pp. 257–282

14. H. Edelsbrunner, J. Harer, *Computational Topology: An Introduction* (American Mathematical Society, New York, 2010)

15. H. Edelsbrunner, D. Letscher, A. Zomorodian, Topological persistence and simplification. Discret. Comput. Geom. **28**(4), 511–533 (2002)

16. E.G. Escolar, Y. Hiraoka, Persistence modules on commutative ladders of finite type. Discret. Comput. Geom. **55**(1), 100–157 (2016)

17. R. Ghrist, Barcodes: the persistent topology of data. Bull. Am. Math. Soc. **45**, 61–75 (2008)

18. A. Hatcher, *Algebraic Topology* (Cambridge University Press, Cambridge, 2002)

19. S. Lang, *Algebra*. Graduate Texts in Mathematics (Springer, Berlin, 2002)

20. D. Letscher, On persistent homotopy, knotted complexes and the alexander module, in *Proceedings of the 3rd Innovations in Theoretical Computer Science Conference, ITCS '12* (ACM, New York, 2012), pp. 428–441

21. S. Mac Lane, *Categories for the Working Mathematician*, vol. 5 (Springer Science & Business Media, Berlin, 2013)

22. N. Milosavljević, D. Morozov, P. Skraba, Zigzag persistent homology in matrix multiplication time, in *Proceedings of the 27th Annual ACM Symposium on Computational Geometry, SoCG '11* (2011), pp. 216–225

23. J.-I. Miyachi, Representations and quivers for ring theorists (2000), http://www.u-gakugei.ac.jp/~miyachi/papers/Yamaguchi.pdf

24. J.R. Munkres, *Elements of Algebraic Topology*. Advanced Book Program (Perseus Books, New York, 1984)

25. S.Y. Oudot, *Persistence Theory: From Quiver Representations to Data Analysis*, vol. 209 (American Mathematical Society, New York, 2015)

26. A. Tausz, G. Carlsson, Applications of zigzag persistence to topological data analysis. CoRR, abs/1108.3545 (2011)

27. The GAP Group, GAP – Groups, Algorithms, and Programming, Version 4.8.7 (2017)

28. A. Zomorodian, *Topology for Computing* (Cambridge University Press, Cambridge, 2005)

29. A. Zomorodian, G. Carlsson, Computing persistent homology, Discret. Comput. Geom. **33**(2), 249–274 (2005)

A Complete Characterization of the One-Dimensional Intrinsic Čech Persistence Diagrams for Metric Graphs

Ellen Gasparovic, Maria Gommel, Emilie Purvine, Radmila Sazdanovic, Bei Wang, Yusu Wang, and Lori Ziegelmeier

Abstract Metric graphs are special types of metric spaces used to model and represent simple, ubiquitous, geometric relations in data such as biological networks, social networks, and road networks. We are interested in giving a qualitative description of metric graphs using topological summaries. In particular, we provide a complete characterization of the one-dimensional intrinsic Čech persistence diagrams for finite metric graphs using persistent homology. Together with complementary results by Adamaszek et al., which imply the results on intrinsic Čech persistence diagrams in all dimensions for a single cycle, our results constitute the important steps toward characterizing intrinsic Čech persistence diagrams for arbitrary finite metric graphs across all dimensions.

E. Gasparovic (✉)
Union College, Schenectady, NY, USA
e-mail: gasparoe@union.edu

M. Gommel
University of Iowa, Iowa City, IA, USA
e-mail: maria-gommel@uiowa.edu

E. Purvine
Pacific Northwest National Laboratory, Seattle, WA, USA
e-mail: emilie.purvine@pnnl.gov

R. Sazdanovic
NC State University, Raleigh, NC, USA
e-mail: rsazdan@ncsu.edu

B. Wang
University of Utah, Salt Lake City, UT, USA
e-mail: beiwang@sci.utah.edu

Y. Wang
The Ohio State University, Columbus, OH, USA
e-mail: yusu@cse.ohio-state.edu

L. Ziegelmeier
Department of Mathematics, Statistics, & Computer Science, Macalester College, Saint Paul, MN, USA
e-mail: lziegel1@macalester.edu

© The Author(s) and the Association for Women in Mathematics 2018

E. W. Chambers et al. (eds.), *Research in Computational Topology*, Association for Women in Mathematics Series 13, https://doi.org/10.1007/978-3-319-89593-2_3

1 Introduction

Graphs are ubiquitous in data analysis, often used to model social, biological, and technological systems. Often, data with a notion of distance can be modeled by a metric graph. A graph is a *metric graph* if each edge is assigned a positive length and if the graph is equipped with a natural metric where the distance between any two points of the graph (not necessarily vertices) is defined to be the minimum length of all paths from one to the other [13]. A metric graph is therefore a special type of metric space that captures simple forms of geometric relations in data that arise in both abstract and practical settings, such as biological networks, social networks, and road networks. For example, the movement patterns that GPS systems trace for vehicles can be modeled as a metric graph for location-aware applications. Brain functional networks as metric graphs capture the blood-oxygen-level dependent signal correlations among different areas of the brain [4]. Social networks as metric graphs can encode strengths of influence between social entities (e.g., persons or corporations). Extracting the topological structures of such networks can provide powerful insights for navigating and understanding their underlying data.

Our work aims to describe topological structures of metric graphs by using persistent homology, a fundamental tool in topological data analysis that has been used in many applications to measure and compare topological features of shapes and functions [10]. In this work, we give a qualitative description of information that can be captured from metric graphs using topological, persistence-based summaries. Theorem 1.1, the main theorem in this paper, provides a complete characterization of the persistence diagrams in dimension 1 for metric graphs in a particular intrinsic setting.

Theorem 1.1 *Let G be a finite metric graph of genus g with shortest system of loops $\{c_1^*, \ldots, c_g^*\}$, and for each $i = 1, \ldots, g$, let $|c_i^*| = \ell_i$ be the length of the ith loop, with $\ell_i \leq \ell_j$ for all $i \leq j$. Then, the one-dimensional intrinsic Čech persistence diagram of G, denoted $\mathrm{Dg}_1 IC_G$, consists of the following collection of points on the y-axis:*

$$\mathrm{Dg}_1 IC_G = \left\{ \left(0, \frac{\ell_i}{4} \right) : 1 \leq i \leq g \right\}.$$

Terms in the statement of this main theorem, including *intrinsic Čech persistence diagram* and *shortest system of loops*, will be rigorously defined in Sects. 2 and 3.1. Intuitively, this theorem provides a way to count and measure the minimal cycles in a metric graph using persistent homology. The one-dimensional intrinsic Čech persistence diagram serves as a kind of fingerprint for the graph's cycle set. Note also that the use of the word *genus* is not referring to the more traditional genus of a surface in which the graph can be embedded. Instead, we use genus to mean the number of cycles in a minimal generating set of the homology of the graph. This will be made more precise in Sect. 3 when we define the shortest system of loops.

Related Work The work of Adamaszek et al. [2, 3] is most relevant to ours, as it helps to characterize persistence diagrams in all dimensions for a metric graph consisting of a single loop. In [3], the authors show that the intrinsic Vietoris–Rips or Čech complex of n points in the circle $\mathbb{S}^1$, at any scale r, is homotopy equivalent to either a point, an odd-dimensional sphere, or a wedge sum of spheres of the same even dimension. The results in [3] further imply that the one-dimensional homology group of a metric graph with a single loop is either rank 1 (in the case where the associated intrinsic complex is homotopy equivalent to $\mathbb{S}^1$) or rank 0 (in all other cases). One can then show that the one-dimensional persistence diagram consists of the single point $\left(0, \dfrac{\ell}{4}\right)$ or $\left(0, \dfrac{\ell}{6}\right)$ in the case of the Čech or Vietoris–Rips filtration, respectively, where ℓ is the length of the loop [2]. (Note: here, we are assuming the convention that the Vietoris–Rips complex at scale r contains all simplices of diameter at most $2r$. This definition differs by a factor of 2 from that used in [2].)

In this paper, we generalize the above result in [2] from a metric graph with a single loop to a metric graph containing an arbitrary, finite set of loops in homological dimension 1. This characterization of persistence diagrams in dimension 1 of an arbitrary finite metric graph complements the work in [2] and constitutes an important step toward the characterization of the intrinsic Čech persistence diagrams of *arbitrary* metric graphs across *all* dimensions.

In addition to the Čech and Vietoris–Rips complexes, there are a number of other types of complexes or combinatorial structures related to graphs. In [16], the author studies the relationship between properties of a graph G and the homology of an associated neighborhood complex. The paper [18] contains a study of the so-called devoid complexes of graphs where simplices correspond to vertex sets whose induced subgraphs do not contain certain forbidden subgraphs. However, the neighborhood and devoid complexes are more related to structural, rather than metric, properties of graphs, so we turn our attention in the remainder of this paper to the more metric-derived Čech complex.

Outline The outline of the paper is as follows. In Sect. 2, we recall the necessary background on persistent homology, in particular for the case that the underlying topological space is a metric graph. Section 3 focuses on establishing the fundamental details of the relationship between a metric graph and its associated intrinsic Čech complex. We prove Theorem 1.1 in Sect. 4. Finally, we discuss our results and plans for future work in Sect. 5.

2 Background

2.1 Homology

Homology is an invariant that characterizes the properties of a topological space X. In particular, the k-dimensional holes (connected components, loops, trapped volumes, etc.) of a space generate a homology group, $H_k(X)$. The rank of this group is referred to as the *k-th Betti number* β_k and counts the number of k-dimensional holes of X. We provide a brief overview of simplicial homology below. For a comprehensive study, see [12, 15]. For a more categorical viewpoint, see [17], and for a discussion of cubical complexes, see [14]. We also note that singular homology is a related concept, and in fact is isomorphic to simplicial homology on spaces which can be triangulated [12]. Although a priori one must work with singular homology of a metric graph G, rather than simplicial homology, metric graphs can be triangulated using a subdivision or discretization of the edges. Therefore, we will work with simplicial homology in the remainder of this paper.

A *simplicial complex* S is a set consisting of a finite collection of k-simplices, where a *k-simplex*, given by $\sigma = [v_0, v_1, \ldots, v_k]$, is the convex hull of distinct points $v_0, v_1, \ldots, v_k$. Thus, a 0-simplex is a vertex, a 1-simplex is an edge, a 2-simplex is a filled-in triangle, a 3-simplex is a solid tetrahedron, and so on. The k-simplices must satisfy the following: (1) if σ is a simplex in S, then all lower-dimensional subsets of σ, called *subsimplices*, are also in S; and (2) two simplices are either disjoint or intersect in a lower-dimensional simplex contained in S.

An algebraic structure of a *vector space* or an *R-module* over some ring R is imposed on the simplicial complex S to uncover the homology of the underlying topological space as follows. The k-simplices form a basis for a vector space, $S^{(k)}$, over some ground field (or ring) $\mathbb{F}$. We call the vector space $S^{(k)}$ the *k-dimensional chain group* over the simplicial complex S. The finite field $\mathbb{Z}_p$ (where p is a small prime), $\mathbb{Z}$, and $\mathbb{Q}$ are common choices for the ground field or ring. In this paper, we will work over $\mathbb{Z}_2$. Furthermore, for each pair of consecutive vector spaces there is a linear map, $\delta_k : S^{(k)} \to S^{(k-1)}$, turning the sequence of chain groups into a *chain complex*:

$$\cdots \to S^{(k+1)} \xrightarrow{\delta_{k+1}} S^{(k)} \xrightarrow{\delta_k} S^{(k-1)} \cdots .$$

These maps are known as *boundary operators*, taking each k-simplex to an alternating sum of its $(k-1)$-subsimplices, its boundary. More precisely, if $[v_0, v_1, \ldots, v_k]$ is a k-simplex, the boundary map $\delta_k : S^{(k)} \to S^{(k-1)}$ is defined by

$$\delta_k([v_0, v_1, \ldots, v_k]) = \sum_{i=0}^{k} (-1)^i [v_0, \ldots, \hat{v}_i, \ldots, v_k]$$

where $[v_0, \ldots, \hat{v}_i, \ldots, v_k]$ is the $(k-1)$-simplex obtained from $[v_0, \ldots, v_k]$ by removing vertex v_i.

The simplicial homology, $H_k(S)$, of a simplicial complex S is defined based on two subspaces of the vector space $S^{(k)}$: $Z_k = ker(\delta_k)$ known as *k-cycles*, and $B_k = im(\delta_{k+1}) = \delta_{k+1}(S^{(k+1)})$ known as *k-boundaries*. Since the boundary operator satisfies the property $\delta_k \circ \delta_{k+1} = 0$ for every $0 \le k \le \dim(S)$, the set of k-boundaries is contained in the set of k-cycles. Then, $H_k(S) = Z_k/B_k$ consists of the equivalence classes of k-cycles that are not k-boundaries (up to *homotopy*).

The elements of $H_k(S)$ are called homology classes and can thus be thought of as equivalence classes represented by cycles enclosing kth order holes that differ by elements of a boundary. The rank of the associated homology group $H_k(S)$ is the number of distinct k dimensional holes and is referred to as the kth Betti number, denoted β_k.

2.2 Persistent Homology and Metric Graphs

In *persistent homology*, rather than studying the topological structure of a single space, X, one considers how the homology changes over an increasing sequence of subspaces. Given a topological space X and a filtration of this space,

$$X_1 \subseteq X_2 \subseteq X_3 \subseteq \ldots \subseteq X_m = X,$$

applying the homology functor gives a sequence of homology groups, called a *persistence module*, induced by inclusion of the filtration

$$H_k(X_1) \to H_k(X_2) \to \ldots \to H_k(X_m).$$

The arrows above indicate the homomorphisms between homology groups induced by the filtration of the space X.

A filtration of a topological space X may be defined in a number of ways. By considering a continuous function $f : X \to \mathbb{R}$, one may define the *sublevel set filtration*

$$f^{-1}(-\infty, a_1) \subseteq f^{-1}(-\infty, a_2) \subseteq \ldots \subseteq f^{-1}(-\infty, \infty),$$

where $a_1 \leq a_2 \leq \ldots \in \mathbb{R}$. Another approach is to build a sequence of simplicial complexes on a set of points using, for instance, the *Vietoris–Rips filtration* [11] or the *intrinsic Čech filtration* [6] discussed below. *Persistent homology* [5, 10] then tracks the elements of each homology group through the filtration. This information may be displayed in a *persistence diagram* for each homology dimension k. A persistence diagram is a set of points in the plane together with an infinite number of points along the diagonal where each point (x, y) corresponds to a homological element that appears (is "born") at $H_k(X_x)$ and which no longer remains ("dies") at $H_k(X_y)$. See Fig. 1 for an example persistence diagram. Distinct topological features may have the same birth and death coordinates; therefore, a persistence diagram is actually a multiset of points. Since all topological features die after they are born, this is an embedding into the upper half plane above the diagonal $y = x$. Points near the diagonal are often considered noise while those further from the diagonal represent more robust topological features. For a detailed description of applications of persistent homology to various problems in the experimental sciences, see [5, 8, 11].

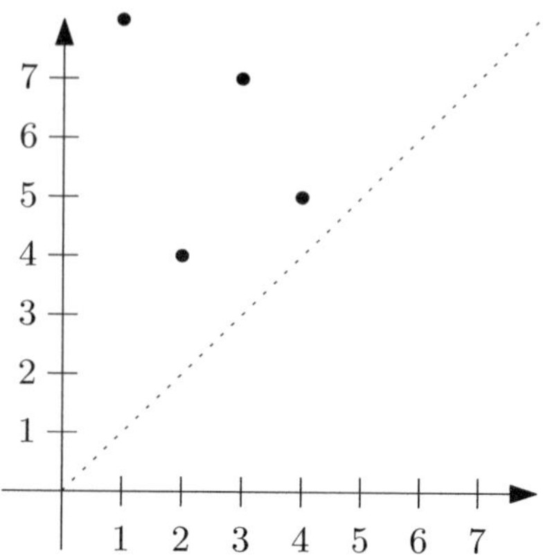

Fig. 1 An example persistence diagram with four points: $(1, 8)$, $(2, 4)$, $(3, 7)$, and $(4, 5)$ corresponding to the birth and death values for distinct topological features

In this paper, we focus on understanding the topological structure of a finite *metric graph* in homology dimension $k = 1$. Given a graph $G = (V, E)$, we define a finite metric graph to be a metric space $(|G|, d_G)$ that is homeomorphic to a one-dimensional finite stratified space consisting of a finite number of 0-dimensional pieces (i.e., vertices) and one-dimensional pieces (i.e., edges or loops) glued together, as described in [1, 9]. More formally, any graph G with vertex set V and edge set E, together with a length function, $length : E \to \mathbb{R}_{\geq 0}$, on E that assigns lengths to edges in E, gives rise to a metric graph $(|G|, d_G)$ where $|G|$ is a geometric realization of G and d_G is defined in the following manner. Using the notation of [9], let e denote an edge in E with $|e|$ being its image in $|G|$, let $e : [0, length(e)] \to |e|$ be the arclength parametrization, and define $d_G(x, y) = |e^{-1}(y) - e^{-1}(x)|$ (with the bars indicating the absolute value) for any $x, y \in |e|$. Now, this definition of $d_G(x, y)$ enables one to define the length of any given path between two points in $|G|$ by first restricting the path to edges in G and then summing the lengths. Then, one may define the distance $d_G(u, v)$ between any pair of points $u, v \in |G|$ to be the minimum length of any path in $|G|$ between u and v. In Fig. 2, we show a graph side-by-side with its corresponding metric graph. Note that in the metric graph we have removed the vertices to emphasize that *all* points along the edges are vertices.

We consider a simplicial complex built on a metric graph as follows. Let (G, d_G) be a metric graph with geometric realization $|G|$. For any point $x \in |G|$, we define the set $B(x, \epsilon) := \{y \in |G| : d_G(x, y) < \epsilon\}$, and we let $U_\epsilon := \{B(x, \epsilon) : x \in |G|\}$ be an open cover. Recall that the set $|G|$ consists of all vertices and every point along an edge, an uncountable set of points. Therefore, U_ϵ is an uncountable cover. The *nerve* of a family of sets $(Y_i)_{i \in I}$ is the abstract simplicial complex defined on the vertex set I by the rule that a finite set $\sigma \subseteq I$ is in the nerve if and only if $\bigcap_{i \in \sigma} Y_i \neq \emptyset$. We let C_ϵ denote the nerve of U_ϵ, referred to as the *intrinsic Čech complex*. In Fig. 3, we show an example subset of U_ϵ and subcomplex of C_ϵ for the metric graph shown in Fig. 2b. We illustrate a finite number of $B(x, \epsilon)$ intrinsic balls and their corresponding nerve.

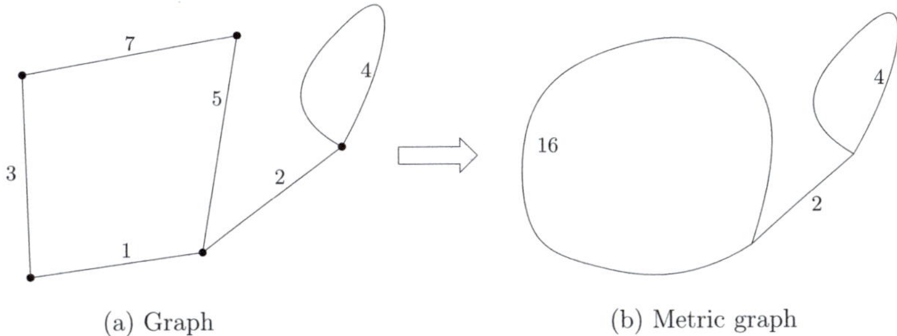

(a) Graph   (b) Metric graph

Fig. 2 A traditional graph with 5 vertices and 6 edges (including one self-loop) alongside its corresponding metric graph

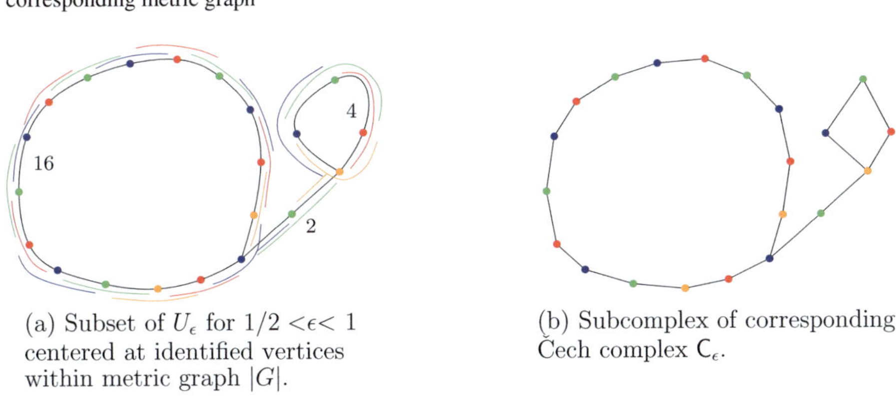

(a) Subset of U_ϵ for $1/2 < \epsilon < 1$ centered at identified vertices within metric graph $|G|$.

(b) Subcomplex of corresponding Čech complex C_ϵ.

Fig. 3 An example cover and corresponding nerve which we have colored to distinguish between overlapping balls and to help illustrate the correspondence between the balls and their associated vertex in the nerve complex

Let C_ϵ^k denote its k-skeleton, where there is a bijection between the vertices in C_ϵ^0 and the points in G. The associated *intrinsic Čech filtration* is defined as the set of inclusion maps

$$\{\mathsf{C}_\epsilon \hookrightarrow \mathsf{C}_{\epsilon'}\}_{\forall 0 \leq \epsilon \leq \epsilon'}.$$

The intrinsic Čech filtration on the metric graph G induces the persistence module

$$\{H_k(\mathsf{C}_\epsilon) \to H_k(\mathsf{C}_{\epsilon'})\}_{\forall 0 \leq \epsilon \leq \epsilon'}$$

in any dimension k, from which we obtain the *intrinsic Čech persistence diagrams*, denoted by $\mathrm{Dg}_k IC_G$. In this paper, we shall only be interested in $\mathrm{Dg}_1 IC_G$. Finally, given an intrinsic Čech complex C_δ, we denote its k-dimensional chain group as $\mathsf{C}_\delta^{(k)}$.

3 From Graphs to Intrinsic Čech Complexes

3.1 Overview and Relevant Notations

We first introduce a few technical definitions and results that will enable us to prove Theorem 1.1. These results will provide a formal way of thinking of properties of a graph and its corresponding Čech complex C_δ, for a sufficiently small δ. This will be achieved in four steps:

- Formal definitions of a shortest system of loops in a graph, shortest-path metric on C_δ (Lemma 1), shortest basis of $H_1(C_\delta)$, and δ-discretization of a graph.
- The graph G has the same homotopy type as its associated intrinsic Čech complex C_δ for a sufficiently small $\delta > 0$ (Lemma 2).
- The inclusion of any δ-discretization $\hat{G}$ of the graph G into C_δ induces an isomorphism between their first homology groups (Lemma 3). This in turn connects a shortest system of loops for G to a basis for the first homology group $H_1(C_\delta)$ of the simplicial complex C_δ.
- The shortest basis of $H_1(C_\delta)$ corresponds to the shortest system of loops for G (Lemma 4).

We tackle the first step in this subsection and the last three steps in Sect. 3.2.

Shortest System of Loops of G For the graph G, we will initially consider its singular homology. In particular, a singular 1-simplex of G is a continuous map $\sigma : [0, 1] \to G$. A singular 1-chain (thus 1-cycle) is a formal sum of such continuous maps [12]. The objects of interest to us are actually the "geometric representations" of such 1-cycles, i.e., their images in G. In particular, a *loop c* in G is a continuous map $c : \mathbb{S}^1 \to G$; we may also use "loop" to refer to the image of this map. (Note that we use the term "loops" in this manner for graphs in order to contrast with "cycles" in homology.) For any singular 1-cycle $\alpha = \sigma_1 + \cdots + \sigma_s$, there is a corresponding loop c whose image in G coincides with the disjoint union of images $\sigma_i([0, 1])$, for $i \in [1, s]$, and we refer to c as the *carrier of α*. All singular 1-cycles carried by the same loop are homologous. For a genus g metric graph G, a *system of loops of G* refers to a set of g loops $\{c_1, \ldots, c_g\}$ such that the homology classes carried by them form a basis for the one-dimensional (singular) homology group $H_1(G)$. For simplicity, we say that these loops are *independent*. Given a system of loops of G, its *length-sequence* is the sequence of lengths of elements in this set in nondecreasing order. A system of loops forms a *shortest system of loops of G* if its length-sequence is smallest in lexicographical order among all systems of loops of G.

From now on, let $\{c_1^*, \ldots, c_g^*\}$ denote a shortest system of loops of G with length-sequence $\ell_1 \leq \ell_2 \leq \ldots \leq \ell_g$, where ℓ_i is the length of the loop c_i for $i = 1, \ldots, g$.

Shortest-Path Distance Metric on C_δ To prove Theorem 1.1, we will only operate on intrinsic Čech complexes. We now define a metric structure $(C_\delta^0, d_{C_\delta})$ on the vertex set C_δ^0 of the Čech complex C_δ for sufficiently small δ.

First, we note that there is a natural bijection between points in G and vertices of C_δ^0. Specifically, for any point $x \in G$, there is a vertex $\overline{x}$ in the nerve complex C_δ corresponding to the covering element (δ-ball) $B(x, \delta)$. In what follows, we say that x *generates* the vertex $\overline{x}$ in the Čech complex C_δ.

Recall that d_G is the shortest-path distance metric on G, where $d_G(x, y)$ is the length of a shortest path between points x and y in G. Given two vertices $\overline{x}, \overline{y} \in C_\delta^0$, we say that a 1-chain γ in $C_\delta^{(1)}$ *connects* $\overline{x}$ with $\overline{y}$ if γ can be represented as

$$\gamma = \sum_{i=0}^{N} [\overline{v}_i, \overline{v}_{i+1}] \text{ such that } \overline{v}_0 = \overline{x}, \; \overline{v}_{N+1} = \overline{y}, \text{ and } [\overline{v}_i, \overline{v}_{i+1}] \text{ are edges in } C_\delta^{(1)}.$$

For simplicity, we also refer to a 1-(simplicial) chain in C_δ as a *path* in C_δ.

Definition 1 We define the metric $d_{C_\delta} : C_\delta^0 \times C_\delta^0 \to \mathbb{R}$ as follows. Given two vertices $\overline{x}, \overline{y} \in C_\delta^0$, if $\overline{x}$ and $\overline{y}$ are connected by an edge in $C_\delta^{(1)}$, then $d_{C_\delta}(\overline{x}, \overline{y}) := d_G(x, y)$. Otherwise, we set $d_{C_\delta}(\overline{x}, \overline{y}) := \inf_\gamma length(\gamma)$, where γ ranges over all 1-chains in C_δ connecting $\overline{x}$ and $\overline{y}$, and its length is defined as

$$length(\gamma) = \sum_{[\overline{v}, \overline{v}'] \in \gamma} d_{C_\delta}(\overline{v}, \overline{v}').$$

In other words, an elementary 1-chain (i.e., an edge) directly inherits the metric from the graph, and the distance between two arbitrary vertices $\overline{x}, \overline{y}$ in C_δ^0 is the length of the "shortest" 1-chain among all 1-chains connecting $\overline{x}$ with $\overline{y}$. However, note that as $d_{C_\delta}(\overline{x}, \overline{y})$ is defined to be the infimum of the lengths of paths connecting $\overline{x}$ with $\overline{y}$, a priori, it is not clear that it can be realized by a shortest path from $\overline{x}$ to $\overline{y}$. We prove the following result, whose proof also implies that there is always a realizing shortest path whose length equals $d_{C_\delta}(\overline{x}, \overline{y})$ for any two $\overline{x}, \overline{y} \in C_\delta^0$.

Lemma 1 *For any two vertices $\overline{x}, \overline{y} \in C_\delta^0$, $d_{C_\delta}(\overline{x}, \overline{y}) = d_G(x, y)$, and the distance $d_{C_\delta}(\overline{x}, \overline{y})$ is realized by a shortest 1-chain connecting $\overline{x}$ to $\overline{y}$.*

Proof We prove by contradiction. (i) First, we assume $d_{C_\delta}(\overline{x}, \overline{y}) < d_G(x, y)$. By definition, there must exist a path $\gamma = \sum_{i=0}^{N} [\overline{v}_i, \overline{v}_{i+1}]$ in $C_\delta^{(1)}$ such that $d_{C_\delta}(\overline{x}, \overline{y}) \leq length(\gamma) < d_G(x, y)$. That is,

$$\sum_{i=0}^{N} d_{C_\delta}(\overline{v}_i, \overline{v}_{i+1}) = \sum_{i=0}^{N} d_G(v_i, v_{i+1}) < d_G(x, y).$$

Therefore, the points v_i in G form a path connecting x and y of shorter length than the shortest path between x, y in G, a contradiction.

(ii) Now assume $d_{C_\delta}(\overline{x}, \overline{y}) > d_G(x, y)$. Suppose ξ is a shortest path connecting x and y in G. Then, we can consider a discrete version of ξ, denoted as $\hat{\xi}$, such that $\hat{\xi}$ contains a finite number of vertices in ξ and each edge is of length at most δ. The vertices and edges in $\hat{\xi}$ give rise to a 1-chain in $\mathbf{C}_\delta^{(1)}$ connecting $\overline{x}$ and $\overline{y}$ that, at the same time, is shorter than γ (the shortest 1-chain connecting $\overline{x}$ with $\overline{y}$). This is a contradiction.

Putting these two directions together, we have that $d_{\mathbf{C}_\delta}(\overline{x}, \overline{y}) = d_G(x, y)$.

Finally, note that by part (ii) above, any shortest path ξ between x and y in G gives rise to a 1-chain $\hat{\xi}$ whose length is at most $d_G(x, y)$. Since we know that $d_{\mathbf{C}_\delta}(\overline{x}, \overline{y}) = d_G(x, y)$, it follows that $length(\hat{\xi}) = d_G(x, y)$. Hence, there exists a shortest path connecting $\overline{x}$ with $\overline{y}$ whose length realizes $d_{\mathbf{C}_\delta}(\overline{x}, \overline{y})$. $\qquad\square$

Shortest Basis of $H_1(\mathbf{C}_\delta)$ The existence of the above metric on the vertex set of $\mathbf{C}_\delta$ allows one to define the *shortest basis* of $H_1(\mathbf{C}_\delta)$, which relates the algebraic construction of $\mathbf{C}_\delta$ to the combinatorial properties of the graph.

Definition 2 Given a homology class $[h] \in H_1(\mathbf{C}_\delta)$, the *length of* $[h]$ is the shortest length of any 1-cycle of $\mathbf{C}_\delta$ contained in this class. We call the corresponding minimal length cycle γ representing $[h]$ the *minimal generating cycle for* $[h]$.

The *shortest homology basis of* $H_1(\mathbf{C}_\delta)$ consists of $\{[\gamma_i]\}_{i=1}^{g}$ such that the (nondecreasing) length-sequence of the basis elements is lexicographically smallest among that of all bases of $H_1(\mathbf{C}_\delta)$. The set of corresponding minimal generating cycles $\{\gamma_1, \dots, \gamma_g\}$ is referred to as the *shortest system of generating cycles for* $H_1(\mathbf{C}_\delta)$ (or the *shortest basis of* $H_1(\mathbf{C}_\delta)$ for short).

In Sect. 3.2, we establish the correspondence between the graph-theoretic notion of the *shortest system of loops* and the topological object *shortest basis of* $H_1(\mathbf{C}_\delta)$ defined above.

δ-Discretization of G Since we work with simplicial homology for $\mathbf{C}_\delta$, we introduce a δ-*discretization of* G, denoted $\widehat{G}$, and consider the simplicial homology of $\widehat{G}$ and its relation to the simplicial homology of $\mathbf{C}_\delta$, instead of the singular homology for G. A δ-discretization is a *subdivision* of G, to use terminology from graph theory, with additional restrictions. For each arc $e = [x, y] \in G$, we replace it with a path of edges $[x_0, x_1] + [x_1, x_2] + \cdots + [x_{k-1}, x_k]$, where $x = x_0$, and $y = x_k$, to obtain $\widehat{G}$ so that: (i) all graph nodes in G are nodes in $\widehat{G}$; (ii) all edges in $\widehat{G}$ are of length at most δ; and (iii) if $length(e) = \ell$, then $\sum_{i=0}^{k-1} length([x_i, x_{i+1}]) = \ell$. In Fig. 3a, the chosen vertices and edges can be seen as a 1-discretization of $|G|$. It is easy to see that $\widehat{G}$ forms a triangulation of G (G is the underlying space of $\widehat{G}$) and thus the (simplicial) homology of $\widehat{G}$ is isomorphic to the (singular) homology of G. Furthermore, since no new loops were created, and all original arc lengths were preserved via a path of shorter edges in the subdivision, a shortest system of loops of G induces a shortest system of loops of $\widehat{G}$ of the same lengths. Hence, from now on, we sometimes abuse the notation slightly and use $\{c_1^*, \dots, c_g^*\}$ to refer to a shortest system of loops in both G and $\widehat{G}$.

3.2 Relating Graphs to Intrinsic Čech Complexes

In this subsection, we formalize the relations between metric graphs and their associated intrinsic Čech complexes. In particular, we provide justification for the intuition that the shortest loops in a graph correspond to the "smallest" basis for the first persistent homology, i.e., shortest loops can be thought of as shortest cycles.

Lemma 2 *Let $\delta < \frac{\ell_1}{4}$, where ℓ_1 is the length of the shortest loop in the shortest system of loops for the graph G. Then, $\mathbf{C}_\delta$ is homotopy equivalent to G.*

Proof We will use the nerve lemma [12] which states that the nerve of a *good cover* of a space X is homotopy equivalent to the space X. A good cover is one in which the intersection of any finite collection of sets in the cover is either empty or contractible. We will prove by contradiction that for a sufficiently small δ, the cover U_δ is a good cover. We first show that any intersection must contain one connected component, and then show that this component must be contractible.

Assume by way of contradiction, $\hat{U}_\delta = \{B(x_1, \delta), \ldots, B(x_k, \delta)\}$ is some finite collection of sets in U_δ such that $B(x_1, \delta) \cap B(x_2, \delta) \cap \cdots \cap B(x_k, \delta)$ contains at least two connected components. Let U and V be two of the connected components contained in this intersection, and consider points $p \in U$ and $q \in V$. Since p and q are in all $B(x_i, \delta)$, there must be a path, π_i, within $|G|$ from p to q in each $B(x_i, \delta)$. Notice that we can choose the π_i to be simple paths such that the length of each π_i is less than 2δ. Additionally, there must be at least two of these paths which are not identical. If all paths are identically equal, then this path would have to be contained in the intersection, contradicting the fact that p and q are in disjoint connected components within the intersection. Without loss of generality, say π_1 in $B(x_1, \delta)$ and π_2 in $B(x_2, \delta)$ are the two distinct paths.

It is possible that these two paths are not entirely disjoint, but because they are not identical, we can find a portion of each which is unique to that path. Travel along both π_1 and π_2 from p until the paths diverge for the first time at point $p' \in |G|$. Then, continue to travel along both until the paths join back again for the first time at point $q' \in |G|$. Let π_1' be the portion of π_1 between p' and q', and similarly π_2' is the portion of π_2 between p' and q'. Now, it is clear that π_1' and π_2' are disjoint except for their endpoints. See Fig. 4 for an illustration. Moreover, the lengths of π_1' and π_2' must be less than 2δ since they are paths contained within two δ balls. Notice that we have created a cycle by taking π_1' from p' to q' and then π_2' from q' back to p'. This cycle has length $length(\pi_1') + length(\pi_2') < 2 \cdot 2\delta < \ell_1$. But, this is a contradiction because we assumed that ℓ_1 is the length of the shortest loop in the shortest system of loops of G, and we have discovered a cycle with length less than ℓ_1. Thus, for any finite collection of sets $\hat{U}_\delta$ in the cover U_δ, the intersection must only contain one connected component.

Now, we must show this component is contractible. Since our underlying space is a metric graph, notice that a connected component must either be a metric tree, or must contain one or more loops. A metric tree is simply connected, hence contractible, so there is nothing to prove in this case. We claim that the other case,

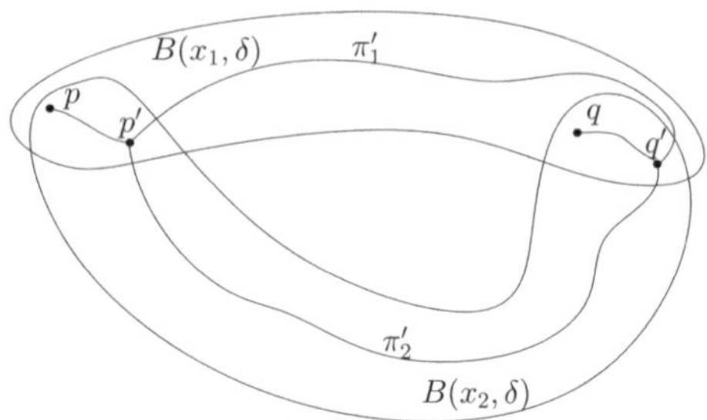

Fig. 4 An illustration of the points and paths described in Lemma 2

where the component contains a loop, cannot happen. If the component did contain a loop, the loop would then be contained entirely inside a δ-ball, and thus would have length less than 2δ. But, by our choice of δ, this is impossible since 2δ is smaller than the length of the shortest loop. We can therefore conclude that the component is contractible, and hence, U_δ is a good cover. By the nerve lemma, it follows that C_δ, the nerve of U_δ, is homotopy equivalent to G. $\qquad\qquad\square$

Before we state and prove the next result, note that there is a natural inclusion map $\iota : \widehat{G} \hookrightarrow C_\delta$. (Note: this map is defined on $\widehat{G}$ rather than G, as it would not be continuous otherwise.) Indeed, each vertex $x \in \widehat{G}$, as we described earlier, is identified with a unique vertex $\iota(x) := \overline{x} \in C_\delta^0$ corresponding to the covering element $B(x, \delta)$. Given an edge $[x, y] \in \widehat{G}$, as $d_G(x, y) \le \delta$ (by construction of the δ-discretization), $B(x, \delta) \cap B(y, \delta) \ne \emptyset$. This means that $[\overline{x}, \overline{y}] \in C_\delta$.

Recall that G is isomorphic to the underlying space of $\widehat{G}$; for simplicity, we identify G with the underlying space of $\widehat{G}$. Given any vertex $\overline{x} \in C_\delta^0$, the point $x \in G$ that generates it may not be a vertex in $\widehat{G}$, in which case there is a unique edge $\sigma \in \widehat{G}$ whose underlying space $|\sigma| \subset G$ contains x. We say that the edge $\iota(\sigma) \in C_\delta^{(1)}$ *covers* $\overline{x}$ (resp. σ *covers* x) in this case.

Finally, given a 1-chain γ of C_δ with only one connected component, we can write it in the form of an ordered sequence of edges: $\gamma = [\overline{x}_1, \overline{x}_2] + [\overline{x}_2, \overline{x}_3] + \cdots + [\overline{x}_{k-1}, \overline{x}_k]$, where each $[\overline{x}_i, \overline{x}_{i+1}]$ is an edge (1-simplex) in C_δ. Note that it is possible that $\overline{x}_i = \overline{x}_j$ for some $i \ne j$. To emphasize this representation of the 1-chain, or path, γ, we also represent it by an ordered sequence of vertices $\langle \overline{x}_1, \overline{x}_2, \ldots, \overline{x}_k \rangle$. A *sub-path* of γ is simply the 1-chain represented by a subsequence $\langle \overline{x}_i, \overline{x}_{i+1}, \ldots, \overline{x}_j \rangle$. For a cycle γ satisfying $\overline{x}_k = \overline{x}_1$, a sub-path will be represented by a subsequence from the cyclic sequence $\langle \overline{x}_1, \overline{x}_2, \ldots, \overline{x}_{k-1}, \overline{x}_1 \rangle$. For example, the subsequence $\langle \overline{x}_{k-2}, \overline{x}_{k-1}, \overline{x}_1, \overline{x}_2, \overline{x}_3 \rangle$ represents a sub-path of the cycle γ.

Lemma 3 *Assume $\delta < \ell_1/4$. The inclusion $\iota : \widehat{G} \hookrightarrow C_\delta$ induces an isomorphism $\iota_* : H_1(\widehat{G}) \to H_1(C_\delta)$.*

Proof For $\delta < \ell_1/4$, we know from Lemma 2 that $H_1(G)$ and $H_1(C_\delta)$ are isomorphic, implying that $\mathrm{rank}(H_1(\widehat{G})) = \mathrm{rank}(H_1(C_\delta))$. Thus, if we can show that the mapping $\iota_* : H_1(\widehat{G}) \to H_1(C_\delta)$ is surjective, then the inclusion ι must induce an isomorphism between these groups. In the following, we prove that any 1-cycle γ in C_δ is homologous to the image of some cycle in $\widehat{G}$ under the inclusion ι.

Fig. 5 Illustration for
Sublemma 1: the (**a**) general
and (**b**) degenerate cases,
respectively. The blue edges
and points are in C_δ but not
$\widehat{G}$, and the red are in both

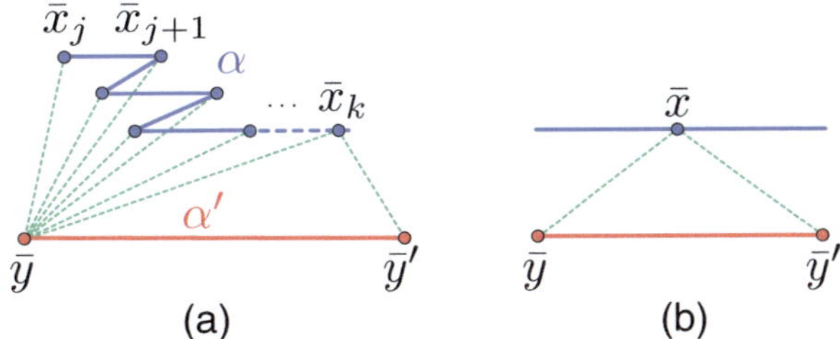

Without loss of generality, we assume that γ has only one connected component
(if it has more than one, we apply the same argument to each of its components).
We view γ as a path $\gamma = [\overline{x}_1, \overline{x}_2] + [\overline{x}_2, \overline{x}_3] + \ldots + [\overline{x}_s, \overline{x}_1]$.

In order to prove the lemma, we require the following two technical sublemmas.
In them, we take a divide-and-conquer approach by decomposing 1-chains in C_δ
and the images of 1-chains in $\widehat{G}$ under ι into pieces, and discuss the relations among
these pieces.

Sublemma 1 *Given an edge $\alpha' = [\overline{y}, \overline{y}']$ in $\iota(\widehat{G})$, where*

$$\alpha = [\overline{x}_j, \overline{x}_{j+1}] + [\overline{x}_{j+1}, \overline{x}_{j+2}] + \cdots + [\overline{x}_{k-1}, \overline{x}_k]$$

is a path in C_δ such that α' covers $\overline{x}_j, \overline{x}_{j+1}, \ldots, \overline{x}_k$, we have

$$\alpha' - \alpha = [\overline{x}_j, \overline{y}] + [\overline{x}_k, \overline{y}'] + \partial[\overline{x}_k, \overline{y}, \overline{y}'] + \partial \sum_{i=j}^{k-1}[\overline{x}_i, \overline{x}_{i+1}, \overline{y}]. \tag{1}$$

*The triangles on the right-hand side of (1) belong to C_δ, so that α' is homologous
to $\alpha + [\overline{x}_j, \overline{y}] + [\overline{x}_k, \overline{y}']$.*

Proof (Sublemma 1) See the illustration in Fig. 5. Since each x_i is covered by the
edge $[y, y'] \in \widehat{G}$, it follows that $d_G(x_i, y), d_G(x_i, y'), d_G(x_i, x_{i+1}) \leq \delta$ for any
$i \in [j, k]$. It then follows that $B(x_k, \delta) \cap B(y, \delta) \cap B(y', \delta) \neq \emptyset$ and $B(x_i, \delta) \cap
B(x_{i+1}, \delta) \cap B(y, \delta) \neq \emptyset$. Hence, the triangles $[\overline{x}_k, \overline{y}, \overline{y}']$ and $[\overline{x}_i, \overline{x}_{i+1}, \overline{y}]$ all belong
to C_δ. Note that this holds for the degenerate case as well, where α' is homologous
to $[\overline{x}, \overline{y}] + [\overline{x}, \overline{y}']$. $\qquad\square$

Sublemma 2 *Given an elementary 1-chain $\alpha = [\overline{x}, \overline{x}']$ in C_δ, consider the shortest
path ξ in G connecting x and x'. Assume that ξ contains a non-empty set of vertices
$y_1, y_2, \cdots, y_l$ in $\widehat{G}$, ordered by their positions in ξ. Consider a 1-chain α' in C_δ of
the form*

$$\alpha' = [\overline{y}_1, \overline{y}_2] + [\overline{y}_2, \overline{y}_3] + \cdots + [\overline{y}_{l-1}, \overline{y}_l].$$

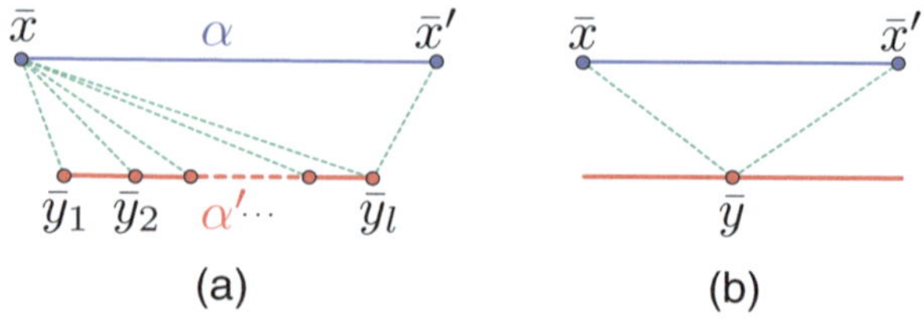

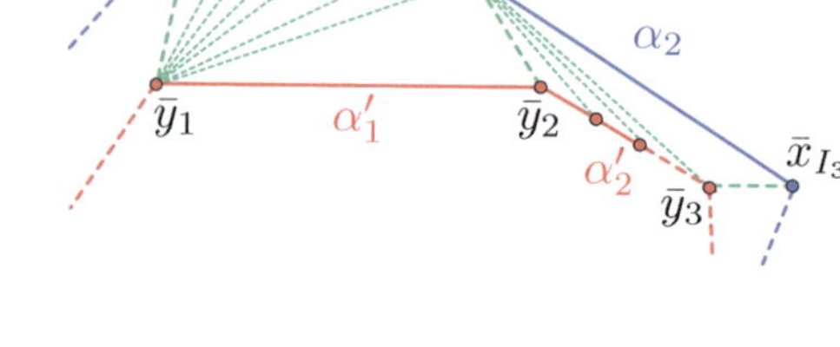

Fig. 6 Illustration of the (**a**) general and (**b**) degenerate cases in Sublemma 2. As before, the blue edges and points are in C_δ but not $\widehat{G}$, and the red are in both

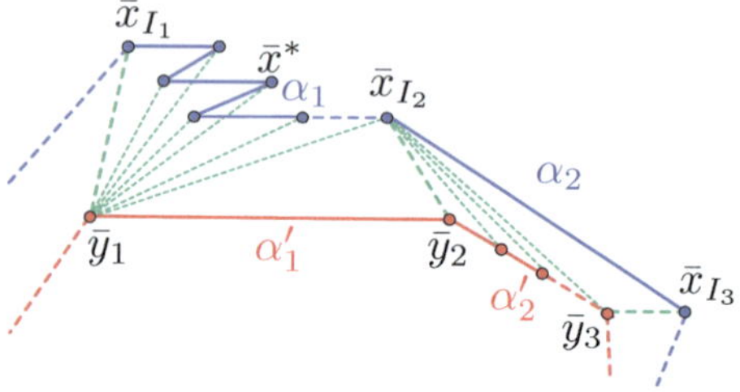

Fig. 7 Illustration for Lemma 3

Obviously, α' is contained in the image $\iota(\widehat{G})$ of $\widehat{G}$ in C_δ, as each $[y_i, y_{i+1}] \in \widehat{G}$ for $i \in [1, l]$. Then,

$$\alpha - \alpha' = [\overline{x}, \overline{y}_1] + [\overline{x}', \overline{y}_l] + \partial[\overline{y}_l, \overline{x}, \overline{x}'] + \partial \sum_{i=1}^{k-1} [\overline{y}_i, \overline{y}_{i+1}, \overline{x}], \qquad (2)$$

where all the triangles on the right-hand side belong to C_δ. This implies that $\alpha' - \alpha$ is homologous to $[\overline{x}, \overline{y}_1] + [\overline{x}', \overline{y}_l]$, and thus α is homologous to $\alpha' + [\overline{x}, \overline{y}_1] + [\overline{x}', \overline{y}_l]$.

Proof (Sublemma 2) See the illustration in Fig. 6. Note that since $[\overline{x}, \overline{x}']$ is an edge in C_δ, $length(\xi) = d_G(x, x') \le 2\delta$. Let z be the midpoint of ξ; obviously, we know $d_G(x, z), d_G(x', z) \le \delta$. Since each $y_i \in \xi$, we have $d_G(y_i, z) \le \delta$ as well for $i \in [1, l]$. It then follows that $z \in B(x, \delta) \cap B(y_i, \delta) \cap B(x', \delta) \ne \emptyset$ and $z \in B(y_i, \delta) \cap B(y_{i+1}, \delta) \cap B(x, \delta) \ne \emptyset$. Hence, the triangles $[\overline{y}_l, \overline{x}, \overline{x}']$ and $[\overline{y}_i, \overline{y}_{i+1}, \overline{x}]$ all belong to C_δ. Note that the argument holds for the degenerate case shown in Fig. 6b in which case we assume that $[\overline{y}, \overline{y}]$ is a degenerate 1-chain to simplify the presentation of the argument. $\qquad \square$

We now prove our Lemma 3 using the previous two sublemmas. The proof is illustrated in Fig. 7. Given the 1-cycle $\gamma = [\overline{x}_1, \overline{x}_2] + [\overline{x}_2 + \overline{x}_3] + \cdots + [\overline{x}_s, \overline{x}_1]$ in C_δ as described earlier, we start with a vertex $\overline{x}^*$ in γ (say $\overline{x}^* = \overline{x}_1$). Without loss of generality, suppose $\overline{x}^*$ is contained in some edge $\alpha'_1 = [\overline{y}_1, \overline{y}_2] \in \iota(\widehat{G})$. We extend in the forward and backward direction along γ to find the maximum sub-path α_1 of γ whose vertices are contained in the edge α'_1. (This is the situation described in Sublemma 1.) Denote the starting vertex of α_1 as $\overline{x}_{I_1}$ and the ending vertex as $\overline{x}_{I_2}$. Now consider the next edge $\alpha_2 = [\overline{x}_{I_2}, \overline{x}_{I_3}]$ in γ where $I_3 \equiv I_{2+1 \bmod s}$. By construction of α_1, we know that x_{I_2} and x_{I_3} are necessarily covered by different edges in $\widehat{G}$. We then construct a path $\alpha_2' \subset \iota(\widehat{G})$ by the procedure described in Sublemma 2. Suppose α_2' is represented by the ordered sequence of vertices $\overline{y}_1', \ldots, \overline{y}_l'$. Note that it is necessary that $\overline{y}_1' = \overline{y}_2$ (which is one of the endpoints of edge α_1').

We then repeat this process. As we traverse γ, we alternate between the situations in Sublemma 1 and Sublemma 2. Therefore, the 1-cycle γ can be represented as a linear combination of 1-chains

$$\gamma = \sum_i^k \alpha_i.$$

Set $\gamma' = \sum_i \alpha_i'$. We argue that γ' is a 1-cycle. Indeed, by construction, the last vertex in the path representation of α_i' is necessarily the same as the first vertex in the path representation of $\alpha_{i+1 \bmod k}$. It then follows that

$$\gamma' - \gamma = \sum_i (\alpha_i' - \alpha_i). \tag{3}$$

Now, combining the claims in Sublemmas 1 and 2, we see that all of the single edges when moved to the right-hand sides of Eqs. (1) and (2) are cancelled out (under $\mathbb{Z}_2$ coefficients) in Eq. (3), leaving only triangles. This is illustrated in Fig. 7, where all edges $[\overline{x}_i, \overline{y}_i]$ cancel while the triangles remain. All of the remaining triangles belong to $\mathbf{C}_\delta$. Hence, $\gamma' - \gamma$ bounds a 2-chain, and thus γ and γ' are homologous. $\qquad\square$

Lemma 4 *Let $\{[\gamma_i]\}_{i=1}^g$ be a shortest homology basis for $H_1(\mathbf{C}_\delta)$ with $\{\gamma_i\}_{i=1}^g$ being the corresponding shortest system of generating cycles, sorted in nondecreasing order of their lengths. Then, its length-sequence equals that of the shortest system of loops of G.*

Proof As before, let $\{c_1^*, \ldots, c_g^*\}$ be a shortest system of loops of $\widehat{G}$ with length sequence $\ell_1 \leq \ell_2 \leq \ldots \leq \ell_g$. Note that this is the same length-sequence of a shortest system of loops for G, as $\widehat{G}$ is a triangulation of G.

First, note that by Lemma 3, $\{[\iota(c_i^*)]\}_{i=1}^g$ form a basis for $H_1(\mathbf{C}_\delta)$. By applying Lemma 1 to each edge in c_i^*, we see that $length(\iota(c_i^*)) = length(c_i^*)$, and thus we know that the length-sequence induced by $\{[\gamma_i]\}_{i=1}^g$ must be lexicographically smaller than or equal to $\{\ell_1, \ell_2, \ldots, \ell_g\}$.

Next, we prove the other direction. Specifically, set $\ell_i' = length(\gamma_i)$ for $i \in [1, g]$. We will prove by contradiction that the length-sequence $\{\ell_1', \ell_2', \ldots, \ell_g'\}$ is lexicographically larger than or equal to $\{\ell_1, \ell_2, \ldots, \ell_g\}$. Assume otherwise; that is, $\{\ell_1', \ell_2', \ldots, \ell_g'\}$ is lexicographically smaller than $\{\ell_1, \ell_2, \ldots, \ell_g\}$.

Let $\overline{V}_\Gamma$ denote the set of vertices from all of the γ_i, and let $V_\Gamma = \{v \in G \mid \overline{v} \in \overline{V}_\Gamma\}$ be the set of corresponding generating points in G. This set is finite since each 1-chain γ_i is finite by definition, and there are finitely many generators. We first refine $\widehat{G}$ to $\tilde{G}$ by subdividing edges of $\widehat{G}$ so that all points in V_Γ are now also vertices in $\tilde{G}$. Obviously, $\tilde{G}$ is also a δ-discretization of G, and thus Lemma 3 holds for it as well.

For each $i \in [1, g]$, suppose the 1-cycle γ_i has the form $\sum_{j=1}^{k} [\overline{x}_j, \overline{x}_{j+1}]$ with $\overline{x}_{k+1} = \overline{x}_1$. It is easy to see that γ_i has only one component. If it has more than one, then there exists at least one component whose corresponding homology class is independent of $[\gamma_1], \ldots, [\gamma_{i-1}]$. We can set γ_i to be this component (which is a 1-cycle itself) and obtain a shorter length-sequence, which contradicts the assumption that $\{[\gamma_i]\}$ is a shortest homology basis for $H_1(C_\delta)$.

We now construct a 1-cycle ξ_i from $\tilde{G}$ satisfying the following conditions: (C-1) $length(\xi_i) = length(\gamma_i)$; and (C-2) $[\iota(\xi_i)] = [\gamma_i]$ where $\iota : \tilde{G} \hookrightarrow C_\delta$ is the inclusion map. First, note that by construction of $\tilde{G}$, for each $\overline{x}_j$, there is a vertex x_j from $\tilde{G}$ such that $\iota(x_j) = \overline{x}_j$. For each $[\overline{x}_j, \overline{x}_{j+1}]$, consider the shortest path (1-chain) π_j connecting x_j to x_{j+1} in $\tilde{G}$. We have that

$$length(\pi_j) = d_G(x_j, x_{j+1}) = d_{C_\delta}(\overline{x}_j, \overline{x}_{j+1}) = length([\overline{x}_j, \overline{x}_{j+1}]).$$

Concatenating all such shortest chains gives rise to a 1-cycle ξ_i in $\tilde{G}$ whose length equals $\sum_{j=1}^{k} length([\overline{x}_j, \overline{x}_{j+1}]) = length(\gamma_i)$. Hence, condition (C-1) above is satisfied.

Furthermore, recall the 1-chain π_j in $\tilde{G}$ corresponding to edge $[\overline{x}_j, \overline{x}_{j+1}]$ constructed above. Suppose π_j is represented by the ordered sequence of vertices $\langle x_j = y_1, y_2, \ldots, y_s = x_{j+1} \rangle$. We claim that the 1-chain $\iota(\pi_j)$ is homotopic to the edge $[\overline{x}_j, \overline{x}_{j+1}]$ in C_δ. Indeed, all vertices from π_j are contained in a path of length $d_G(x_j, x_{j+1}) \leq 2\delta$ (as $[\overline{x}_j, \overline{x}_{j+1}]$ is an edge in C_δ). It then follows that $B(x_j, \delta) \cap B(y_a, \delta) \cap B(y_{a+1}, \delta) \neq \emptyset$ for any $a \in [1, s-1]$ (the common intersection contains, say, the midpoint of path π_j). Hence, triangles of the form $[\overline{x}_j, \overline{y}_a, \overline{y}_{a+1}]$, for $a \in [1, s-1]$, exist in C_δ, establishing a homotopy between $\iota(\pi_j) = \langle \overline{x}_j, \overline{y}_2, \ldots, \overline{y}_s = \overline{x}_{i+1} \rangle$ and the edge $[\overline{x}_j, \overline{x}_{j+1}]$ in C_δ. All of these homotopies together, for all $i \in [1, k]$, provide a homotopy between the 1-cycle $\iota(\xi_i)$ and γ_i. Thus, $[\iota(\xi_i)] = [\gamma_i]$ and condition (C-2) above also holds.

Hence, from $\{\gamma_i\}_{i=1}^{g}$, we can obtain a set of 1-cycles $\{\xi_i\}_{i=1}^{g}$ of $\tilde{G}$ whose length-sequence equals $\{\ell_1', \ldots, \ell_g'\}$. Furthermore, by Lemma 3 and condition (C-2) above, $\{[\xi_i]\}_{i=1}^{g}$ must form a basis for $H_1(\tilde{G})$ as $\{[\iota(\xi_i)]\}_{i=1}^{g} = \{[\iota(\xi_i)]\}_{i=1}^{g}$ form a basis for $H_1(C_\delta)$. It then follows that we have obtained a system of loops $\{\xi_i\}_{i=1}^{g}$ for $\tilde{G}$ whose length-sequence is smaller than $\{\ell_1, \ldots, \ell_k\}$, which contradicts the fact that the latter is the shortest length-sequence possible. Hence, our assumption is wrong, and the length-sequence for $\{\gamma_i\}_{i=1}^{g}$ cannot be smaller than $\{\ell_1, \ldots, \ell_k\}$.

Therefore, putting the proofs of both directions together, we have that the length-sequence for the shortest homology basis $\{[\gamma_i]\}_{i=1}^g$ must be equal to that of the shortest system of loops for $\tilde{G}$ and thus, for G. $\qquad\square$

4 Proof of Main Theorem

We are now ready to prove Theorem 1.1. Let $\mu_\epsilon : \mathbf{C}_\delta \to \mathbf{C}_\epsilon$ for $\epsilon > \delta$ denote the chain map given by inclusion, and let $\mu_\epsilon^c : \mathbf{C}_\delta^{(1)} \to \mathbf{C}_\epsilon^{(1)}$ denote the associated inclusion map of one-dimensional chain groups. The latter induces the map on one-dimensional homology $\mu_\epsilon^h : H_1(\mathbf{C}_\delta) \to H_1(\mathbf{C}_\epsilon)$. Let $\gamma \in \mathbf{C}_\epsilon^{(1)}$ denote a cycle, with $[\gamma] \in H_1(\mathbf{C}_\epsilon)$ the corresponding homology class.

Proof (Proof of Theorem 1.1) Let $\{[\gamma_i]\}_{i=1}^g$ be a shortest homology basis for $H_1(\mathbf{C}_\delta)$ corresponding to the shortest system of loops $\{c_1^*, \ldots, c_g^*\}$ of G. First, note that $\mu_\epsilon^c(\gamma_i)$ is a boundary cycle in $\mathbf{C}_\epsilon$ for $\epsilon = \dfrac{\ell_i}{4}$. This is due to the fact that, for any triple of points $x, y, z \in \gamma_i$, $B\left(x, \dfrac{\ell_i}{4}\right) \bigcap B\left(y, \dfrac{\ell_i}{4}\right) \bigcap B\left(z, \dfrac{\ell_i}{4}\right) \neq \emptyset$. Therefore, γ_i must die at $\dfrac{\ell_i}{4}$ or earlier. The rest of the proof consists of showing that:

A) For $i = 1, \ldots, g$, the ith cycle does not die before $\epsilon = \dfrac{\ell_i}{4}$; and

B) No other cycles are created due to interference between cycles.

Notice that A) and B) can be reformulated to the language of bases, where condition A) is equivalent to a linear independence condition and B) is equivalent to a spanning condition. Therefore, the proof of Theorem 1.1 follows from Proposition 4.1 below. $\qquad\square$

Proposition 4.1 *For any $i = 1, \ldots, g$, the set*

$$\{[\mu_\epsilon^c(\gamma_i)], [\mu_\epsilon^c(\gamma_{i+1})], \ldots, [\mu_\epsilon^c(\gamma_g)]\}$$

is a basis for $H_1(\mathbf{C}_\epsilon)$ where $\dfrac{\ell_{i-1}}{4} \leq \epsilon < \dfrac{\ell_i}{4}$ and $\ell_0 = 0$.

Proof (Proposition 4.1) We will prove the two conditions A) and B).

For A) the linear independence condition, we show that $\displaystyle\sum_{j=i}^g a_j [\mu_\epsilon^c(\gamma_j)] = [0]$ implies coefficients $a_j = 0$ for all j. Let $\gamma = \displaystyle\sum_{j=i}^g a_j \mu_\epsilon^c(\gamma_j)$ be a cycle representing the trivial class $[0] = [\gamma] \in H_1(\mathbf{C}_\epsilon)$. Assume, by way of contradiction, that there exists j with $i \leq j \leq g$ such that $a_j \neq 0$. Since $[\gamma] = [0]$, there exists a two-dimensional chain $\alpha \in C_\epsilon$ having γ as its boundary, i.e., $\partial\alpha = \gamma$. Let $\alpha = \sum_{k \in J} \Delta_k$ where, for some index set J, $\{\Delta_k\}_{k \in J}$ is the set of 2-simplices in the triangulation of α, and where for each k, $t_k := \partial\Delta_k \in C_\epsilon^{(1)}$. Then, $\gamma = \partial\alpha = \partial \sum_k \Delta_k = \sum_k \partial\Delta_k = \sum_k t_k$, that is,

$$\gamma = \sum_{j=i}^{g} a_j \mu_\epsilon^c(\gamma_j) = \sum_k t_k. \tag{4}$$

We aim to contradict the fact that some $a_j \neq 0$ in the above sum.

To this end, we define a map $\rho : C_\epsilon^{(1)} \to C_\delta^{(1)}$ for $\epsilon > \delta$ by specifying its effect on edges in the Čech complex C_ϵ and extending the map linearly to all 1-chains in $C_\epsilon^{(1)}$. First, there is a natural bijection between the set of vertices of C_δ and that of C_ϵ; specifically, the vertex in C_δ representing the δ-ball $B(u, \delta)$ corresponds to the vertex in C_ϵ representing the covering element $B(u, \epsilon)$. For simplicity, we assume $C_\delta^0 = C_\epsilon^0$ and use $\bar{u}$ to denote the vertex in C_δ (resp. in C_ϵ) representing the covering element $B(u, \delta)$ (resp. $B(u, \epsilon)$). Now, given an edge $e = [\bar{u}, \bar{v}] \in C_\epsilon^{(1)}$, we describe its image $\rho(e)$ in $C_\delta^{(1)}$. If $[\bar{u}, \bar{v}]$ spans an edge in C_δ, then we set $\rho(e)$ to be that edge. Otherwise, the existence of the edge $[\bar{u}, \bar{v}] \in C_\epsilon^{(1)}$ implies, by definition, that $B(u, \epsilon) \cap B(v, \epsilon) \neq \emptyset$. Choose an arbitrary (but fixed with respect to $\bar{u}$ and $\bar{v}$) point $w \in B(u, \epsilon) \cap B(v, \epsilon) \subseteq G$, and define $\rho(e)$ to be the concatenation of a shortest 1-chain in $C_\delta^{(1)}$ between $\bar{u}$ and $\bar{w}$ and a shortest 1-chain between $\bar{w}$ and $\bar{v}$. We claim that the length of the 1-chain $\rho(e)$ is at most 2ϵ. Indeed, by construction and Lemma 1, we have:

$$\text{length}(\rho(e)) = d_{C_\delta}(\bar{u}, \bar{w}) + d_{C_\delta}(\bar{w}, \bar{v}) = d_G(u, w) + d_G(w, v) \leq \epsilon + \epsilon = 2\epsilon. \tag{5}$$

Notice that the restriction $\rho|_{C_\delta^{(1)}} : C_\delta^{(1)} \to C_\delta^{(1)}$ is the identity mapping. Clearly, $\rho|_{C_\delta^{(1)}}$ is the identity on the basis elements, the edges in the Čech complex $C_\delta^{(1)}$, since there is no shorter path within $C_\delta^{(1)}$ than the edge itself. Then, by linearity, $\rho|_{C_\delta^{(1)}}$ is the identity on all of $C_\delta^{(1)}$. Additionally, $\rho(\mu_\epsilon^c(\gamma_j)) = \rho(\gamma_j) = \gamma_j$ since $\mu_\epsilon^c(\gamma_j) = \gamma_j \in C_\epsilon^{(1)}$. Applying ρ to Eq. (4) we obtain the following:

$$\rho(\gamma) = \sum_{j=i}^{g} a_j \gamma_j = \sum_k \rho(t_k). \tag{6}$$

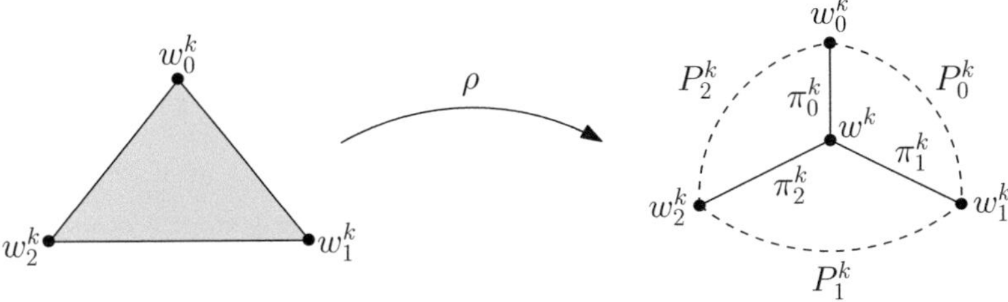

Fig. 8 Action of ρ on the triangle $t_k = [w_0^k, w_1^k, w_2^k]$. Notice the three cycles contained in $\rho(t_k)$

Next, we show that for each k, $\rho(t_k)$ is the sum of short cycles. Notice that $t_k = [w_0^k, w_1^k, w_2^k] = \partial \Delta_k$ represents a trivial cycle in $\mathbf{C}_\epsilon^{(1)}$, so there must exist some point $w^k \in \bigcap_{n=0}^{2} B(w_n^k, \epsilon)$. See Fig. 8.

Consequently, for $n = 0, 1, 2$, there exist the following paths π_n^k and P_n^k in the Čech complex $\mathbf{C}_\delta^{(1)}$:

- $\pi_n^k = [w^k, w_n^k]$, which has length less than or equal to ϵ, and
- $P_n^k = \rho([w_n^k, w_{(n+1 \mod 3)}^k])$, which has length at most 2ϵ by (5).

Therefore, $\rho(t_k) = \sum_{n=0}^{2}[\pi_n^k + P_n^k - \pi_{(n+1 \mod 3)}^k]$ is the sum of three cycles, each of length at most $\epsilon + \epsilon + 2\epsilon = 4\epsilon$. Since the length of each smaller cycle is less than 4ϵ, $\rho(t_k)$ can be expressed in terms of the shortest basis $\{\gamma_j\}_{j=1}^{i-1}$ for $H_1(\mathbf{C}_\delta)$:

$$\rho(\gamma) \;=\; \sum_{k} \rho(t_k) = \sum_{k}\sum_{j=1}^{i-1} a_j^k \gamma_j = \sum_{j=1}^{i-1} a_j' \gamma_j. \tag{7}$$

$$\Longrightarrow \sum_{j=i}^{g} a_j \gamma_j \overset{(6)}{=} \sum_{j=1}^{i-1} a_j' \gamma_j \tag{8}$$

$$\Longrightarrow \sum_{j=1}^{i-1} a_j' \gamma_j + \sum_{j=i}^{g}(-a_j)\gamma_j = 0. \tag{9}$$

As the set $\{\gamma_i\}_{i=1}^{g}$ is a shortest basis for $H_1(\mathbf{C}_\delta)$, the coefficients in the above sums must all be zero, that is $a_j = 0$ for all j, which contradicts our initial assumption. Therefore, the set $\{[\mu_\epsilon^c(\gamma_i)], [\mu_\epsilon^c(\gamma_{i+1})], \ldots, [\mu_\epsilon^c(\gamma_g)]\}$ is linearly independent in $H_1(\mathbf{C}_\epsilon)$. In particular, γ_i does not become trivial before $\dfrac{\ell_i}{4}$.

Next, to prove B), we show that the map $\mu_\epsilon^h : H_1(\mathbf{C}_\delta) \to H_1(\mathbf{C}_\epsilon)$ is surjective by showing that it has a right inverse up to homotopy. In particular, we will show

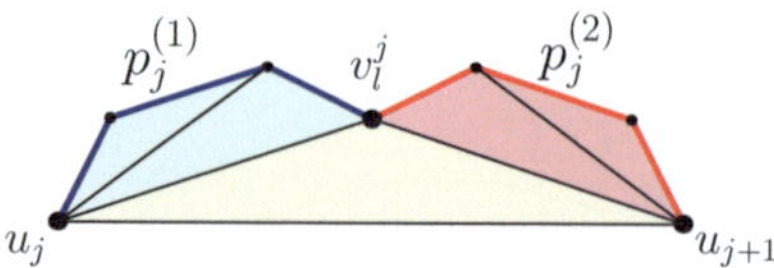

Fig. 9 A part of C_ϵ used to illustrate that $[\rho(\eta)] = [\eta]$. In particular, each edge $[u_j, u_{j+1}]$ will be mapped by ρ to a chain of edges. The path $p_j^{(1)}$ is colored in blue, and $p_j^{(2)}$ is colored in red. The homotopy is realized in two two-dimensional simplices (represented by the blue, red, and yellow shading) that exist in C_ϵ based on the Čech construction

that for every $[\eta] \in H_1(C_\epsilon)$,

$$\mu_\epsilon^h([\rho(\eta)]) = [(\mu_\epsilon^c \circ \rho)(\eta)] = [\eta] \in H_1(C_\epsilon) \tag{10}$$

where the chain $\eta \in C_\epsilon^{(1)}$ is a geometric realization of the class $[\eta]$.

Consider a cycle $\eta = \langle u_0, u_1, \ldots, u_k, u_0 \rangle \in C_\epsilon^{(1)}$ representing $[\eta] \in H_1(C_\epsilon)$. Set $p_j = \rho([u_j, u_{j+1}]) = \langle u_j, v_1^j, \ldots, v_{m_j}^j, u_{j+1} \rangle$, for $j = 0, 1, \ldots, k$ and $u_{k+1} = u_0$. Then, the image $\rho(\eta)$ is just a concatenation of these paths $\rho(\eta) = p_0 + p_1 + \cdots + p_k \in C_\delta^{(1)}$. Since μ_ϵ^c is induced by inclusion, we abuse the notation slightly and use $p_0 + p_1 + \cdots + p_k$ to denote the image $\mu_\epsilon^c(\rho(\eta))$ in $C_\epsilon^{(1)}$ as well.

To show that Eq. (10) holds, we need to prove that $[\mu_\epsilon^c(\rho(\eta))]_{C_\epsilon} = [\eta]_{C_\epsilon}$, which is achieved by showing that p_j is path homotopic to edge $[u_j, u_{j+1}]$ in C_ϵ for all $j = 0, 1, \ldots, k$ and $u_{k+1} = u_0$.

To this end, note that by construction of $\rho([u_j, u_{j+1}])$, the path p_j is (i) either the edge $[u_j, u_{j+1}]$; or (ii) consists of two shortest 1-chains $p_j^{(1)}$ and $p_j^{(2)}$, where $p_j^{(1)}$ and $p_j^{(2)}$ are from u_j to v_l^j and v_l^j to u_{j+1}, respectively (see Fig. 9)—here, v_l^j corresponds to the node $\bar{w}$ in the definition of $\rho(e)$ earlier. If it is case (i), then obviously, p_j is homotopic to edge $[u_j, u_{j+1}]$ (as they are the same). Now, consider case (ii). Note by definition of ρ, $d_{C_\delta}(u_j, v_l^j) \leq \epsilon$ and $d_{C_\delta}(v_l^j, u_{j+1}) \leq \epsilon$. As $p_j^{(1)}$ (resp. $p_j^{(2)}$) is a shortest path between u_j and v_l^j (resp. between v_l^j and u_{j+1}), it then follows that $d_G(u_j, v_n^j) = d_{C_\delta}(u_j, v_n^j) \leq \epsilon$ for all $0 \leq n < l$. Similarly, $d_G(u_{j+1}, v_n^j) \leq \epsilon$ for all $l \leq n < m_n$. Therefore, each of the following is a two-dimensional simplex: the triangle $[u_j, v_n^j, v_{n+1}^j] \in C_\epsilon$ for all $0 \leq n < l$ and the triangle $[u_{j+1}, v_n^j, v_{n+1}^j] \in C_\epsilon$ for all $l \leq n < m_n$. It then follows that the path (1-chain) p_i is homotopic to the edge $[u_j, u_{j+1}]$ for case (ii), as well. Concatenating these homotopies proves the homotopy equivalence of η and $\rho(\eta)$. Hence, $[\mu_\epsilon^c(\rho(\eta))] = [\eta]$ which establishes Eq. (10).

Notice that $[\rho(\eta)] = \sum_{j=i}^{g} a_j[\gamma_j] \in H_1(\mathbf{C}_\delta)$ since $\dfrac{\ell_{i-1}}{4} \leq \epsilon < \dfrac{\ell_i}{4}$. By equation (10), we have

$$[\eta] = \mu_\epsilon^h([\rho(\eta)]) = \mu_\epsilon^h(\sum_{j=i}^{g} a_j[\gamma_j])$$

$$= \sum_{j=i}^{g} a_j[\mu_\epsilon^c(\gamma_j)] \in \mathrm{Span}(\{[\mu_\epsilon^c(\gamma_j)]\}_{j \geq i}),$$

which completes the proof of the surjectivity of μ_ϵ^h. This establishes the spanning condition B). In other words, if $[\eta]$ is a homology class in $H_1(\mathbf{C}_\epsilon)$, then it must be formed only from homology classes $[\mu_\epsilon^c(\gamma_j)]$ for $j \geq i$, and thus no additional cycles are created. $\square$

5   Future Work

The overarching theme of this work is to show how persistence may be used to obtain qualitative–quantitative summaries of metric graphs that reflect the underlying topology of the graphs. We obtained a complete characterization of all possible intrinsic Čech persistence diagrams in homological dimension one for metric graphs. What is currently known regarding the characterization of intrinsic Čech persistence diagrams for metric graphs is summarized in a diagram shown in Fig. 10. The horizontal axis represents the homological dimension and the vertical axis represents the genus of a graph. In this setting, the previous results of Adamaszek et al. [2], who consider the intrinsic Čech persistence diagrams in all dimensions for a graph that consists of a single loop, lie on the horizontal strip at height one, while the results in this paper constitute the blue vertical strip. The rest of the upper-right quadrant is unknown, and our hope is to make further progress toward a complete characterization of the intrinsic Čech persistence diagrams associated to arbitrary metric graphs. Moreover, we aim to generalize our results to the Vietoris–Rips complex.

The choice of a particular complex may be inspired by particular graph features that one is interested in. A *graph motif* is usually thought of as a graph on a small number of vertices (in general, any graph pattern can be a motif). Counting the number of small motifs in a graph is equivalent to the subgraph isomorphism problem, which is NP-complete. Since persistence has a polynomial time computational complexity, the question we would like to answer is: can the intrinsic Čech or other related persistence diagrams be used to determine or approximate graph motif counts? Additionally, the local version of this question, the number of graph motifs incident with a particular vertex, may be approached via the local

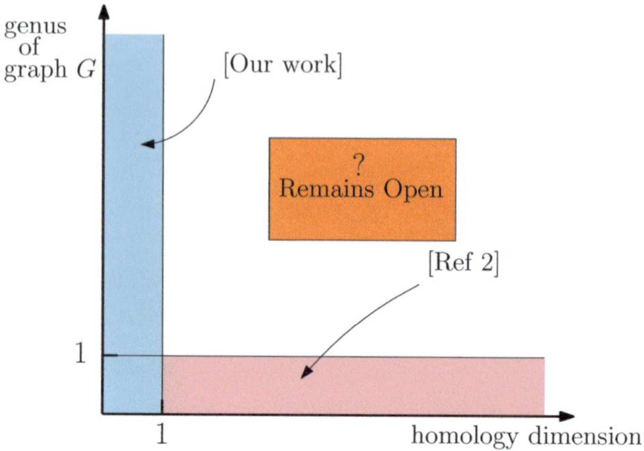

Fig. 10 Figure summarizing the results from this paper and from [2]

homology at a vertex (homology of the k-neighborhood of a vertex relative to its boundary). As a start, persistence-based characterizations of a class of graph motifs should be obtained. Depending on the type of characterization obtained, we would be interested in determining to what extent our persistence-based summaries could be useful in the classification of the motifs present in a query graph.

Ultimately, the complete or partial characterization of the topological information about a graph that is captured by persistent homology associated to various chain complex constructions is closely related to comparing their discriminative powers. In particular, we are interested in comparing the Čech and persistence distortion distance summaries.

The intrinsic Čech filtration and associated persistence diagrams allow one to define the *intrinsic Čech distance*, d_{IC}, between two metric graphs (G_1, d_{G_1}) and (G_2, d_{G_2}). This distance, introduced in [6], is defined as follows:

$$d_{IC}(G_1, G_2) := d_B(Dg_1 IC_{G_1}, Dg_1 IC_{G_2}),$$

where d_B is the bottleneck distance between the two intrinsic Čech persistence diagrams in dimension 1.

The *persistence distortion distance*, d_{PD}, that was first introduced in [9], is more closely related to the metric properties of a graph. Given a base point $s \in |G|$, define $f_s : |G| \to \mathbb{R}$ to be the geodesic distance to the base point s, i.e., $f_s(x) = d_G(s, x)$ for all $x \in |G|$. Then, $Dg_1 f_s$ is the 1st-extended persistence diagram [7] associated to the sublevel set filtration induced by f_s. One may do this for any given base point in the metric graph, yielding a set of persistence diagrams for each graph. Let

$$\phi : |G| \to SpDg$$

$$s \mapsto Dg f_s$$

where $SpDg$ denotes the space of all persistence diagrams. Then, $\phi(|G|) \subset SpDg$ is called the *persistence distortion* of G. The *persistence distortion distance* between two metric graphs is defined to be the Hausdorff distance between their persistence distortion sets:

$$d_{PD}(G_1, G_2) := d_H(\phi(|G_1|), \phi(|G_2|)).$$

A natural question to ask is whether or not d_{PD} is more discriminative than d_{IC}, i.e., whether or not there exists a constant $c > 0$ such that

$$d_{IC}(G_1, G_2) \le c \cdot d_{PD}(G_1, G_2).$$

We are currently working on extending the preliminary results that establish the inequality for certain classes of metric graphs to arbitrary input graphs.

Acknowledgements We are grateful for the Women in Computational Topology (WinCompTop) workshop for initiating our research collaboration. In particular, participant travel support was made possible through an NSF grant (NSF-DMS-1619908), and some additional travel support and social outings throughout the workshop were made possible through a gift from Microsoft Research. The Institute for Mathematics and its Applications generously offered the use of their facilities, their experienced staff to facilitate conference logistics, and refreshments. We appreciate their continued support of the applied algebraic topology community with regard to the WinCompTop Workshop, the special thematic program *Scientific and Engineering Applications of Algebraic Topology* held during the 2013–2014 academic year, and the Applied Algebraic Topology Network WebEx talks. Finally, we would like to thank the AWM ADVANCE grant for travel support for organizers and speakers to attend the WinCompTop special session at the AWM Research Symposium in April 2017.

EP was partially supported by the Asymmetric Resilient Cybersecurity Initiative at Pacific Northwest National Laboratory, part of the Laboratory Directed Research and Development Program at PNNL, a multi-program national laboratory operated by Battelle for the U.S. Department of Energy. During the completion of this project, RS was partially supported by the Simons Collaboration Grant, BW was partially supported by NSF-IIS-1513616, and YW was partially supported by NSF-CCF-1526513 and NSF-CCF-1618247.

References

1. M. Aanjaneya, F. Chazal, D. Chen, M. Glisse, L. Guibas, D. Morozov, Metric graph reconstruction from noisy data. Int. J. Comput. Geom. Appl. **22**(04), 305–325 (2012)
2. M. Adamaszek, H. Adams, The Vietoris-Rips complexes of a circle (2015). arXiv 1503.03669
3. M. Adamaszek, H. Adams, F. Frick, C. Peterson, C. Previte-Johnson, Nerve complexes of circular arcs. Discret. Comput. Geom. **56**(2), 251–273 (2016)
4. B. Biswal, F.Z. Yetkin, V.M. Haughton, J.S. Hyde, Functional connectivity in the motor cortex of resting human brain using echo-planar MRI. Magn. Reson. Med. **34**, 537–541 (1995)
5. G. Carlsson, Topology and data. Bull. Am. Math. Soc. **46**(2), 255–308 (2009)
6. F. Chazal, V. de Silva, S. Oudot, Persistence stability for geometric complexes. Geom. Dedicata. **173**, 193–214 (2014)
7. D. Cohen-Steiner, H. Edelsbrunner, J. Harer, Extending persistence using Poincaré and Lefschetz duality. Found. Comput. Math. **9**(1), 79–103 (2009)

8. V. de Silva, R. Ghrist, Coverage in sensor networks via persistent homology. Algebr. Geom. Topol. **7**, 339–358 (2007)
9. T.K. Dey, D. Shi, Y. Wang, Comparing graphs via persistence distortion, in *Proceedings 31st International Symposium on Computational Geometry*, ed. by L. Arge, J. Pach. Leibniz International Proceedings in Informatics (LIPIcs), vol. 34, pp. 491–506 (Dagstuhl, Wadern, 2015). Schloss Dagstuhl–Leibniz-Zentrum fuer Informatik
10. H. Edelsbrunner, J. Harer, Persistent homology - a survey. Contemp. Math. **453**, 257–282 (2008)
11. R. Ghrist, Barcodes: the persistent topology of data. Bull. Am. Math. Soc. **45**, 61–75 (2008)
12. A. Hatcher, *Algebraic Topology* (Cambridge University Press, Cambridge, 2002)
13. P. Kuchment, Quantum graphs I. Some basic structures. Waves Random Media **14**(1), S107–S128 (2004)
14. W.S. Massey, *A Basic Course in Algebraic Topology*. Graduate Texts in Mathematics (Springer, New York, 1991)
15. J.R. Munkres, *Elements of Algebraic Topology*. Advanced Book Classics (Perseus Books, New York, 1984)
16. C. Previte, The $\mathcal{D}$ -neighborhood complex of a graph. PhD thesis, Colorado State University, 2014
17. J. Rotman, *An Introduction to Homological Algebra*. Universitext (Springer, New York, 2009)
18. D. Taylan, Matching trees for simplicial complexes and homotopy type of devoid complexes of graphs. Order **33**, 459–476 (2016)

Comparing Directed and Weighted Road Maps

Alyson Bittner, Brittany Terese Fasy, Maia Grudzien, Sayonita Ghosh Hajra, Jici Huang, Kristine Pelatt, Courtney Thatcher, Altansuren Tumurbaatar, and Carola Wenk

Abstract With the increasing availability of GPS trajectory data, map construction algorithms have been developed that automatically construct road maps from this data. In order to assess the quality of such (constructed) road maps, the need for meaningful road map comparison algorithms becomes increasingly important.

A. Bittner
Department of Mathematics, University at Buffalo (SUNY), Buffalo, NY, USA
e-mail: alysonbi@buffalo.edu

B. T. Fasy (✉)
Gianforte School of Computing and Department of Mathematical Sciences, Montana State
University, Bozeman, MT, USA
e-mail: brittany@cs.montana.edu

M. Grudzien
School of Computing, Montana State University, Bozeman, MT, USA

S. Ghosh Hajra
Department of Mathematics, Hamline University, St Paul, MN, USA
e-mail: sghoshhajra01@hamline.edu

J. Huang
Gianforte School of Computing, Montana State University, Bozeman, MT, USA
e-mail: jici.huang@msu.montana.edu

K. Pelatt
Department of Mathematics, St. Catherine University, St Paul, MN, USA
e-mail: kepelatt@stkate.edu

C. Thatcher
Department of Mathematics and Computer Science, University of Puget Sound, Tacoma, WA,
USA
e-mail: cthatcher@pugetsound.edu

A. Tumurbaatar
Department of Mathematics and Statistics, Washington State University, Pullman, WA, USA
e-mail: altaa.tumurbaatar@wsu.edu

C. Wenk
Department of Computer Science, Tulane University, New Orleans, LA, USA
e-mail: cwenk@tulane.edu

© The Author(s) and the Association for Women in Mathematics 2018

E. W. Chambers et al. (eds.), *Research in Computational Topology*, Association
for Women in Mathematics Series 13, https://doi.org/10.1007/978-3-319-89593-2_4

Indeed, different approaches for map comparison have been recently proposed; however, most of these approaches assume that the road maps are modeled as undirected embedded planar graphs.

In this paper, we study map comparison algorithms for more realistic models of road maps: directed roads as well as weighted roads. In particular, we address two main questions: how close are the graphs to each other, and how close is the information presented by the graphs (i.e., traffic times, trajectories, and road type)? We propose new road network comparisons and give illustrative examples. Furthermore, our approaches do not only apply to road maps but can be used to compare other kinds of graphs as well.

1 Introduction

Road maps have traditionally been constructed by performing costly and time-consuming land surveys. With increasing availability of GPS trajectory data (e.g., as in [25]), map construction algorithms have been developed that automatically construct road maps from such data sources. In order to assess the quality of such (constructed) road maps, meaningful comparison of different road maps is important. Indeed, different approaches for map comparison have been recently proposed in the literature [1, 2, 7, 20]; however, most of these approaches assume that the road maps are modeled as undirected embedded graphs.

In this paper, we use topology to inform graph comparison. As higher-ordered topological features are not present in graphs and connectivity of graphs is well-studied, we focus on the role of one-cycles (i.e., loops) in graphs. One-cycles play a fundamental role in the combinatorics of the road network (e.g., in defining the options available for navigation). In some cases, we may focus on the one-cycles defined by city blocks; that is, loops in planar embedded graphs that contain no graph edges inside the loop. More generally, we can allow for any set of (directed) one-cycles.

After presenting related work in Sect. 2, we investigate several suitable filtrations on graphs in Sect. 3, and we present our approach for comparing two different graphs in Sect. 4. Our goal is to be able to determine how close the graphs are to each other and to quantify the similarity of information presented in the graphs (such as traffic times or road type).

2 Related Work

When considering graph comparison from a graph-theoretical point of view, the obvious approaches to model the problem are NP-hard, such as subgraph isomorphism or graph edit distance [11, 14, 26]. However, these generally require one-to-one mappings of the graphs and do not take any geometric embedding into account. For comparing road maps, distance measures have been proposed that generally model the maps as embedded undirected graphs; see [3, 4, 7] for surveys as well as experimental comparisons and quality assessments. The (directed or undirected) Hausdorff distance can be used to compare the sets of points covered by each graph [5], but this does not take any structural information of the graphs into account. Other distance measures have been proposed that compare random samples of shortest paths [20, 22] or all paths [2] in the street maps. In order to compare more general topological information, Biagioni and Eriksson developed a sampling-based distance measure [7] and Ahmed et al. introduced the local persistent homology distance [1]. Still, few of these algorithms have theoretical guarantees, and most do not deal with weighted or directed graphs. This paper fills the gap by proposing ways we can compare weighted and directed graphs using topological information found via the intrinsic distance measure in the graph. The information extracted can be local or global in nature; see the swatch filtration in Sect. 3.1 and the AMD filtration in Sect. 3.2, respectively.

2.1 Weighted and Directed Graph Comparison

Any graph comparison approach has to map features from one graph to the other in order to define a distance measure. An example is subgraph isomorphism, which is NP-hard in general but can be solved in polynomial time for planar graphs [16]. A connectivity-based dissimilarity measure is used in [30], where the authors generalize a standard distance metric in graph theory to the context of directed weighted graphs. In the case where there is a bijection ψ between the nodes in the two graphs G_1 and G_2, one can define a metric dependent on ψ by comparing the distance between nodes i, j in G_1 and nodes $\psi(i)$, $\psi(j)$ in G_2. Explicitly, this metric, called the *total distortion*, is defined as follows:

$$d_\psi(G_1, G_2) = \Sigma_{i,j \in G_1} |d_{G_1}(i, j) - d_{G_2}(\psi(i), \psi(j))|$$

To obtain a distance between the graphs not dependent on the bijection, we consider the minimum distance over all bijections ψ:

$$v(G_1, G_2) = \min_\psi d_\psi(G_1, G_2)$$

Since these paths are now taking into account the distance as defined in the directed weighted graph setting, this metric encodes that data as well.

Another approach is to use random walks, as in [21, 29], where two graphs G_1 and G_2 are compared by computing random walks of a fixed length in the product space $G_1 \times G_2$, and then computing the so-called *label sequence kernel* (here, labels are referring to the edge labels), which can formally be expressed as:

$$\kappa(G_1, G_2) := \sum_{h_1} \sum_{h_2} \kappa(h_1, h_2) p(h_1|G_1) p(h'|G'),$$

where $\kappa(h_1, h_2)$ is an inner-product kernel between weighted edges h_1 and h_2 in G_1 and G_2, and $p(h|G)$ is the probability of h being a random walk in G.

2.2 Persistence-Based Comparisons

Persistent homology is a tool that describes the structure of a point cloud at multiple scales (more generally, of a filtered topological space). Moreover, the persistence diagram (a multi-set of birth–death pairs) can be computed in matrix multiplication time using basic linear algebra. Next, we briefly introduce the relevant concepts. For more details on the fundamentals of persistent homology, we refer the reader to [8, 10, 15, 17, 23], and to [19, 24] for algebraic topology.

Intuitively, the zero- and one-dimensional homology groups, denoted as $H_0(\mathbb{X})$ and $H_1(\mathbb{X})$, respectively, correspond to path-connected components and loops of $\mathbb{X}$. (In this paper, we are not interested in two- and higher-dimensional homology, as the data we consider are graphs). In some examples, we work with the *relative homology* groups $H_k(\mathbb{X}, \mathbb{A})$ (for $\mathbb{A} \subseteq \mathbb{X}$), which is equivalent to $H_k(\mathbb{X}/\mathbb{A})$ when $\mathbb{X}$ is a simplicial complex and $\mathbb{A}$ is a subcomplex of $\mathbb{X}$ [19, Prop. A5]. The common setting for persistent homology is when we have a finite sequence of topological spaces connected by inclusions, called a discrete *filtration*. If $\mathbb{X}$ is embedded in $\mathbb{R}^2$, we could, for example, use the height filtration where $\mathbb{X}_t = \{(x, y) \in \mathbb{X} \: s.t. \: y \leq t\}$. Here, t is the filtration parameter and is colloquially referred to as *time*. Applying the homology functor to this sequence, we obtain a sequence of vector spaces connected by homomorphisms $f_k^{i,j} : H_k(\mathbb{X}_i) \to H_k(\mathbb{X}_j)$ for $i \leq j$. Persistent homology tracks the so-called *birth* and *death* events of $\mathbb{X}_t$ as t ranges from $-\infty$ to ∞. These birth and death times can be paired, resulting in a set of birth–death pairs that can be plotted in $\mathbb{R}^2$ in a *persistence diagram*, or plotted as stacked intervals in a *barcode plot*.

For two diagrams $\mathcal{D}_1$ and $\mathcal{D}_2$, a well-defined and commonly used distance is the *bottleneck distance*:

$$W_\infty(\mathcal{D}_1, \mathcal{D}_2) := \inf_{f : \mathcal{D}_1 \to \mathcal{D}_2} \sup_{x \in \mathcal{D}_1} ||x - f(x)||, \tag{1}$$

where f is a bijection between $\mathcal{D}_1$ and $\mathcal{D}_2$ [15, Ch. VII]. This distance between diagrams is Lipschitz-stable both for general filtrations [9, 13] and for the local homology filtrations [1, 6]. More generally, one may use the Wasserstein distance between diagrams, $W_p^p(\mathcal{D}_1, \mathcal{D}_2) := \inf_f \sum_{x \in \mathcal{D}_1} ||x - f(x)||^p$, under which the space of persistence diagrams is complete and separable but not a Hilbert space (which is needed for many machine learning and statistical tools).

For graphs (or any subset of a metric space), one approach is to utilize local persistent homology to generate a distance metric [1]. The *local persistent homology distance* computes bottleneck distance between the local persistence diagrams of G and G' over a common finite open cover $\mathcal{U} = \{U_1, U_2, \ldots, U_n\}$ of the domain of interest (e.g., over disks of a fixed radius centered at lattice points) [1]. Here, the local persistence diagram over U_i, denoted $\mathcal{D}(G, U_i)$, corresponds to the persistence module $\{H_k(G_t, G_t - U_i)\}_t$ and captures the local homology of the road networks as seen from the set U_i. A measure of distance is obtained by taking the weighted average of local distances over $\mathcal{U}$:

$$d_{lph}(G, G') := \sum_{U \in \mathcal{U}} \omega_U W_\infty(\mathcal{D}(G, U), \mathcal{D}(G', U)),$$

where ω_U is a weight function. In Sect. 4.1, we extend this definition to a set of paired persistence diagrams.

Recently, the TDA community has made progress in using a persistence-based approach for comparing weighted digraphs. In [28], Turner defined two filtrations of directed graphs. In particular, geodesic paths are used to define an *ordered tuple (OT) filtration*. Persistent homology with the OT filtration has the same persistent homology as the Rips filtration, when the underlying graph is undirected. In another setting, directed graph homology introduced in [18] has been extended to the persistence setting in [12].

For the comparison of weighted graphs, we introduce the idea to use correspondence between cycles. When comparing different data on the same graph, constructing such correspondences is trivial. In applications that compare road maps, correspondences may be obtained via mappings between city blocks or landmarks. While, as a general theoretical problem, constructing the best correspondence might be difficult, this work is motivated by a special setting: detecting and quantifying changes in road networks. In this case, the graph changes slowly over time in predictable ways (roads opening / closing is common, but roads moving or shifting is uncommon); thus, most underlying road network cycles stay the same.

3 Filtrations

In this section, we introduce three different filtrations from which we can compute persistence diagrams in order to compare two directed and / or weighted graphs in Sect. 4. Some of the filtrations below require an additional parameter choice (e.g., choice of basepoint), which has some choice involved; see Sect. 5.

3.1 Swatch Filtration

Let $G = (V, E)$ be a graph with weight function $f : E \to \mathbb{R}_{\geq 0}$ defined over the edges. To define the *swatch filtration*, we start with a choice of initial vertex (or root) $v \in V$ and define $G_t = G_t(f, v)$ to be the subgraph of G induced by the set of all (possibly directed) paths in G with initial vertex v and length less than or equal to t. The length of a (directed) path $p = (e_1, \ldots, e_n)$ in G with respect to f is defined to be $\ell_f(p) := \Sigma_{i=1}^n f(e_i)$. If G is a directed graph without an accompanying function, we assume that the unit weights and so the length of a directed path is simply the number of edges in that path. We note here that these subgraphs G_t are the so-called *swatches* defined in [27]; we propose using the radius threshold t as a persistence

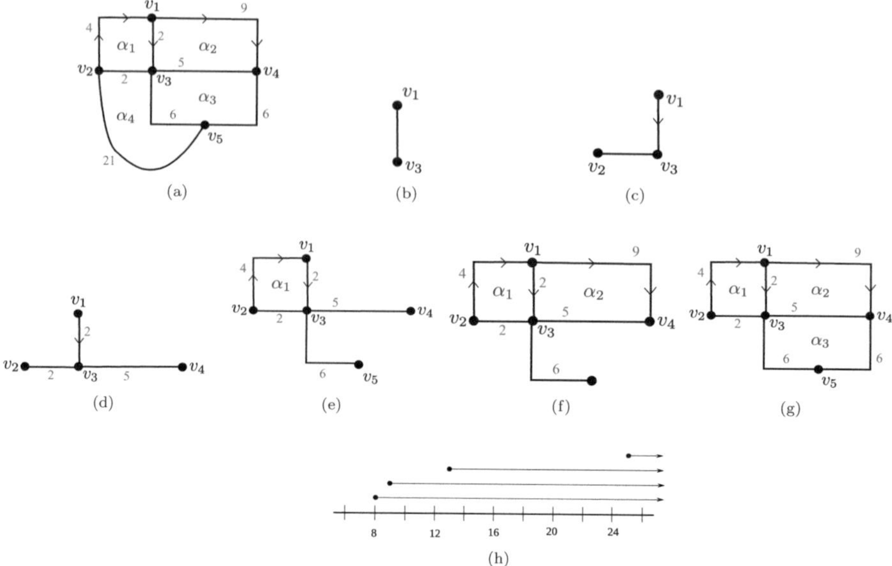

Fig. 1 Example of a swatch filtration of the directed weighted graph G using v_1 as the root node. The faces of the planar graph are labeled with the cycle that bounds the face. Notice that $\cup G_t = G$ for $t > 25$, as the set of all paths of length at most 25 covers G. (**a**) Weighted digraph G. (**b**) $\cup G_t$ for $2 < t < 4$. (**c**) $\cup G_t$ for $4 < t < 7$. (**d**) $\cup G_t$ for $7 < t < 8$. (**e**) $\cup G_t$ for $8 < t < 9$. (**f**) $\cup G_t$ for $9 < t < 13$. (**g**) $\cup G_t$ for $13 < t < 25$. (**h**) Resulting barcode for G

parameter in order to define a persistence barcode for each $v \in V$. We note that this diverges from [27], where histograms of swatches of a fixed radius (called *clothes*) were used to compare dynamical processes.

Example 1 (Swatch Filtration) In Fig. 1, we show an example of the swatch filtration with the choice of v_1 as the initial vertex. Notice that the homology of this graph is generated by the following cycles: α_1, α_2, α_3, and α_4. These cycles are minimal in this embedding, as each has an empty interior. In subfigures (a)–(g), we see the union of paths in G_t, denoted by $\cup G_t$. The first value of t for which the cycle α_1 appears is $t = 8$, so the birth time $b(\alpha_1) = 8$, as shown in the barcode in Fig. 1h. Notice that since no two-cells are introduced, no one-cycles ever die in this filtration.

The swatch diagram has two potential use-cases. First, a natural choice for the initial vertex v_1 might exist (e.g., the vertex represents a landmark such as the main train station or a baseball stadium). In this case, the swatch diagram can be used to compare the road networks *from the perspective of the landmark*. On the other hand, a choice of a landmark is not always clear, and different landmarks can create vastly different swatch filtrations; see Example 2.

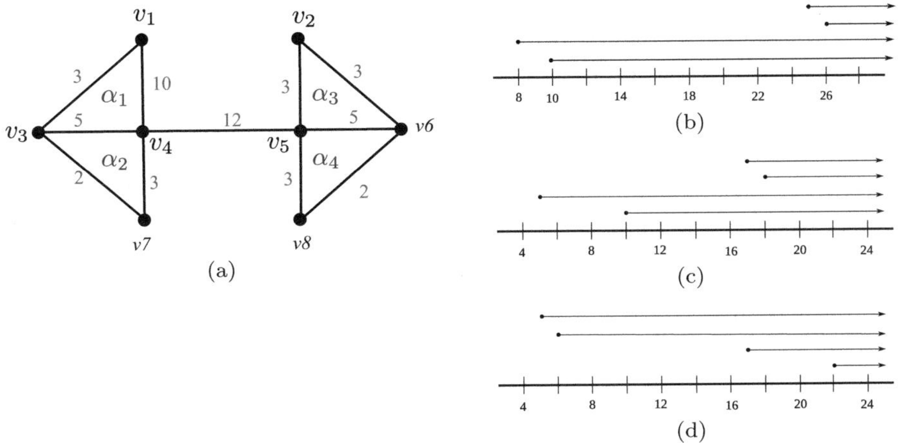

Fig. 2 The swatch filtrations constructed from different initial vertices (basepoints) have different barcodes. The edges are not drawn to scale. (**a**) Undirected graph G with edge weights. (**b**) Barcode using v_1 as initial vertex. (**c**) Barcode using v_4 as initial vertex. (**d**) Barcode using v_5 as initial vertex

Example 2 (Different Basepoints) The birth times of the cycles depend on the choice of initial vertex, which means that the distance based on the swatch filtration is sensitive to choice of basepoint. In this example, we calculate the birth times for the cycles in Fig. 2, starting at three different vertices: v_1, v_4, and v_5, resulting in three different barcodes.

3.2 Average Minimum Distance

To avoid the arbitrary choice of an initial vertex as we encountered above, we consider a second filtration using the average minimum distance (AMD) to each of the other vertices.

Definition 1 (Average Minimum Distance) The graph geodesic distance (or the minimum distance) from vertex v_i to vertex v_j is defined as: $d(v_i, v_j) := \min_p \{\ell(p)\}$, where p ranges over all possible (directed) paths between v_i and v_j. The degree-k AMD, $\mathrm{AMD}_k \colon V \to \mathbb{R}$, of vertex v_i is the average of the minimum distances to all of the other vertices:

$$\mathrm{AMD}_k^k(v_i) := \frac{1}{|V| - 1} \sum_{v_j \neq v_i} (d(v_i, v_j))^k.$$

We define G_0 to be the empty set and G_t to be the subgraph of G induced by the set of all vertices of AMD less than or equal to t.

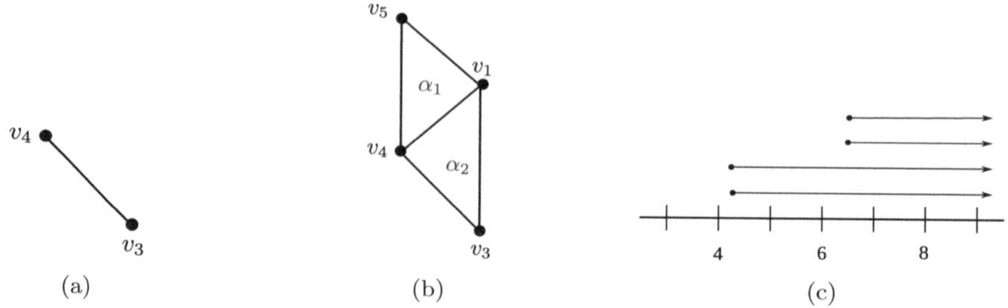

Fig. 3 Example of an average minimum distance (AMD) filtration for the graph found in Fig. 1. We calculate the AMD for v_1 as $(7 + 3 + 3 + 4)/4 = 4.25$, so v_1 appears in G_t when $t \geq 4.25$. Likewise, we have $\text{AMD}_1(v_2) = 6.5$, $\text{AMD}_1(v_3) = 3.0$, $\text{AMD}_1(v_4) = 3.0$, and $\text{AMD}_1(v_5) = 4.25$. Notice that $G_t = G$ for all $t > 6.5$. (**a**) G_t for $3 < t < 4.25$. (**b**) G_t for $4.25 < t < 6.5$. (**c**) Resulting barcode

We note here that AMD_k is related to the eccentricity function:

$$\text{ECC}(v_i) := \frac{1}{|V| - 1} \max_{v_j \neq v_i} d(v_i, v_j) = \lim_{k \to \infty} \text{AMD}_k(v_i).$$

In other words, the eccentricity of vertex v_i is the maximum geodesic distance from v_i to any other vertex in the graph. On the other hand, the function $\text{AMD}_k(v_i)$ is the weighted average of the minimum distances to every other vertex in the graph, and limits to the eccentricity.

Example 3 (AMD Filtration) In Fig. 3, we use the AMD filtration of the graph G from Fig. 1. The AMDs of the vertices in G are $\text{AMD}_1(v_1) = 5.25$, $\text{AMD}_1(v_2) = 5.25$, $\text{AMD}_1(v_3) = 4.75$, $\text{AMD}_1(v_4) = 7.25$, and $\text{AMD}_1(v_5) = 8$.

3.3 Killing Cycles

The filtrations described above do not include higher-dimensional cells; hence, every cycle *lives forever*. In order to decrease the complexity of the homology groups, we need to either delete edges or add two-cells. We construct a cone vertex c at time zero with an edge connecting it to the initial vertex for the swatch filtration (or the first vertex added for the AMD filtration). Then, for each edge $e = (v_i, v_j) \in V \times V$, we add the (unoriented) edges (c, v_i) and (c, v_j) along with the triangle $\Delta = (c, v_i, v_j)$ at the parameter value $\max\{c\, \ell_f(e), \text{appear}(e)\}$, for some constant $c \in R_{\geq 0}$ and where $\text{appear}(e)$ is the first parameter containing e in the filtration. So, the additional simplices we add at parameter t in the filtration are:

$$C_t = \{c\} \cup \{(c, v), (c, w), (c, v, w) \mid (v, w) \in G_t \text{ and } c\, \ell_f(v, w) \leq t\}.$$

Thus, we are considering the filtrations of the form $\{\overline{G}_t := G_t \cup C_t\}$, where G_t is either the swatch or AMD filtration. The effect of adding these simplices is as follows: In the swatch filtration, cycles must be longer for larger values of t in order to be seen from the initial vertex, which intuitively makes sense. In the AMD filtration, the lengths of the cycles in $\overline{G}_t$ will be at least $\frac{2t}{c}$.

4 Comparing Data on Graphs

Given two road networks as directed graphs, both embedded in a compact subset of $\mathbb{R}^2$, we want to define a topology-based distance metric between the two networks. See Sect. 2 for related work and a discussion on the difficulty of this problem. We start with a simplified problem: to develop a method for measuring distance between a graph $G = (V, E)$ with two different annotations, such as two different weight functions on the edges $f, g : E \to \mathbb{R}$ or two different directed graphs with the same underlying undirected graph.

4.1 Paired Analysis

For each $v \in V$, we compute two persistence diagrams $\mathcal{D}_v(f)$ and $\mathcal{D}_v(g)$, corresponding to annotations f and g, respectively. Thus, we can mirror the approach of [1] for aggregating paired diagrams (see Sect. 2) in order to compute a distance between annotations f and g:

$$d_{\text{pair}}(f, g) := \sum_{v \in V} \omega_v W_\infty \left(\mathcal{D}_v(f), \mathcal{D}_v(g) \right),$$

where ω_v is a weight function on the vertices of G, assumed to be the uniform weight $1/|V|$, unless otherwise stated. Since the underlying graphs are exactly the same, we can construct an even more sensitive distance between filtrations $G_t(f, v)$ and $G_t(g, v)$, which we describe next.

4.2 Birth–Birth Diagrams

Using the swatch and AMD filtrations from the previous section, we notice that these filtrations depend only on the intrinsic information on the graphs, whether this information is given by functions (such as f, g) or by direction assignments to the edges. Furthermore, each complex in the swatch and AMD filtrations is a subgraph of G, and all one-cycles have infinite persistence. Rather than looking at persistent homology as the traditional birth–death diagram, we designate a special subset of

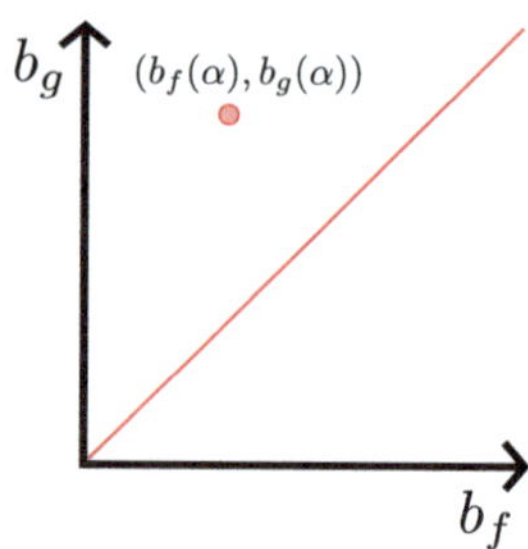

Fig. 4 A birth–birth diagram

cycles $A \subseteq H_1(G)$ of interest (e.g., A could be a set of generating cycles). For each cycle in A, we have two birth times (one for each filtration). We use the two birth times of the cycles to develop a distance measure between the two different annotations, and we visualize this information in a birth–birth diagram; see Fig. 4.

More formally, let $G = (V, E)$ be a directed graph, and let $f, g \colon E \to \mathbb{R}_{\geq 0}$ be two annotations that we wish to compare. We filter the paths p in G by this length $\ell_f(p)$ (or, alternatively, by one of the filtrations defined in the previous section), and say that the minimum distance necessary to complete a cycle $\alpha = (v_1, \ldots, v_n)$ is computed by considering all possible ways to break the cycle into two connected subpaths. More explicitly, we compute this minimum as:

$$b_f(\alpha) := \min_{v_i, v_j \in \alpha} \left\{ \max\{\ell_f(\alpha_{i,j}), \ell_f(\alpha_{j,i})\} \right\},$$

where $\alpha_{i,j}$ is the subpath of α from v_i to v_j. (Using either the swatch or AMD filtration, we can define $b_f(\alpha)$ to be the minimum parameter t such that $\alpha \in G_t$.) We note that $b_f(\alpha)$ is well-defined.

To compare the intrinsic information on a graph G given two annotations $f, g \colon E \to \mathbb{R}$, we begin by choosing a subset of cycles $A \subseteq H_1(G)$; for example, we could use a set of generating cycles. For each $\alpha \in A$, we compute the birth times $b_f(\alpha)$ and $b_g(\alpha)$. Thus, we have two paired sets of birth times for each cycle in A. We visualize this information in what we call the *birth–birth diagram*, where the birth for the filtration corresponding to f is the x-coordinate and for g the y-coordinate; see Fig. 4. We then define the distance between the two annotated graphs to be the total L_∞-distance between points in the birth–birth diagram and the diagonal $y = x$; in other words, we compute

$$D\left((G, f), (G, g)\right) = \sum_{\alpha_i \in A} ||b_f(\alpha_i) - b_g(\alpha_i))||_\infty.$$

We note here that sometimes it may be convenient to first normalize the parameters so that $t = 1$ corresponds to the end of both filtrations.

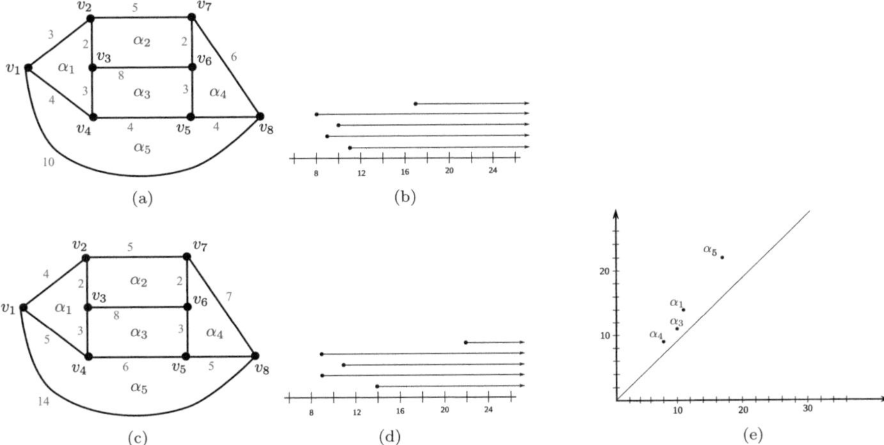

Fig. 5 Example of two swatch filtrations derived from two different weights on the same underlying graph. Here, we use v_6 as the root vertex to construct both filtrations. The distance between these two graphs is $10 = 3 + 0 + 1 + 1 + 5$. (**a**) Weighted graph G. (**b**) Resulting barcode for G. (**c**) Weighted graph G'. (**d**) Resulting barcode for G'. (**e**) Birth–birth diagram

Notice that this construction considers a single graph with two different annotations. These techniques can be extended to analyze two annotations on different graphs (G, f) and (G', f'). To do so, rather than starting with $A \subseteq H_1(G)$, we start with a correspondence $C \subset H_1(G) \times H_1(G')$. Given such a correspondence between cycles, we proceed as above to plot a point in the birth–birth diagram for every $(\alpha, \alpha') \in C$.

4.3 Examples

This section provides enhanced examples to clarify some of the concepts presented above.

Example 4 (Birth–Birth Diagram for Two Weighted Graphs) In Fig. 5, we give an example of two swatch filtrations from the same undirected graph but with different weight functions, and the corresponding birth–birth diagram. Here, the filtrations use v_6 as the root vertex.

Example 5 (Birth–Birth Diagram for Two Directed Graphs) In Fig. 6, we give an example of two filtrations from the same underlying graph with the direction of one edge changed, and the corresponding birth–birth diagram.

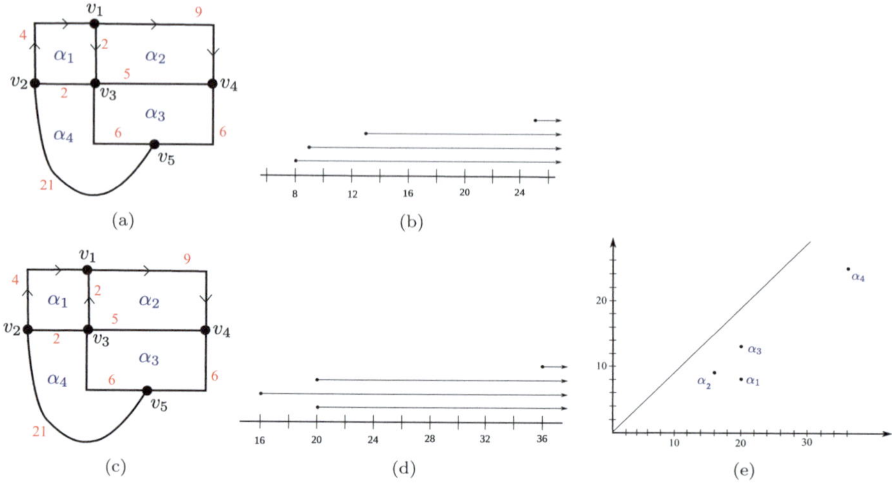

Fig. 6 Example of two swatch filtrations starting at v_1 on two weighted directed graphs, and the resulting birth–birth diagram. The graphs G and G' are identical, except for the direction of the edge (v_1, v_3). (**a**) Weighted digraph G. (**b**) Resulting barcode for G. (**c**) Weighted digraph G'. (**d**) Resulting barcode for G'. (**e**) Birth–birth diagram

5 Discussion

In this paper, we present a framework for comparing weighted and directed graphs. We have introduced new filtrations for directed graphs that capture the complexity of potential paths in a graph by focusing on cycles in the graph. We note that the set of filtrations presented is not exhaustive but is a practical starting point for using topological techniques for comparing weighted directed graphs. Future work will provide an analysis of these techniques on real data sets.

The swatch filtration presented has a persistence module corresponding to each vertex. In addition, if we allow for the killing cycles enhancement (as described in Sect. 3.3), we have an additional parameter c to account for. The parameter c can be optimally chosen by cross-validation; however, a single vertex might not be able to reveal all necessary information through its corresponding swatch filtration. So, we suggest randomly sampling the vertices to have a basepoint set when the set of vertices is too large to consider all possibilities.

The contribution of this paper that we suspect will have the most impact is the birth–birth diagram, which can be used in settings outside of directed, weighted graphs. In fact, it can be applied in any setting where we have multiple functions defined over the same domain (e.g., temperatures at different times, observations before and after a stimulus in a scientific experiment, etc.). Of course, in other settings, the features of interest we wish to match might not be one-cycles.

We note that the theory presented here is not limited to these filtrations. For example, we ask: what topological properties of a given directed graph embedded in $\mathbb{R}^2$ during the filtration can be uncovered? Can we incorporate the embedding rather than just the intrinsic distances? Moreover, empirical study of these filtrations will determine if they can be used in practice to compare different weighted (directed) graphs. As a first step, we will investigate how these methods can be used to compare traffic patterns across large timescales, where the road networks might change over time (bridges added, new exits on highways, etc.). We also note that the assumptions do not require the graphs to be planar; hence, they provide an obvious advantage over most of the existing methods to compare road networks. Analyzing this advantage will be part of our future research.

Acknowledgements This paper is the product of a working group of WinCompTop 2016, sponsored by NSF DMS 1619908, Microsoft Research, and the Institute for Mathematics and Its Applications (IMA) in Minneapolis, MN. In addition, part of this research was conducted under NSF CCF 618605 (Fasy) and NSF CCF 1618469 (Wenk).

References

1. M. Ahmed, B.T. Fasy, C. Wenk, Local persistent homology based distance between maps, in *Proceedings of 22nd ACM SIGSPATIAL International Conference on Advances in Geographic Information Systems* (ACM, New York, 2014), pp. 43–52
2. M. Ahmed, B.T. Fasy, K.S. Hickmann, C. Wenk, Path-based distance for street map comparison. ACM Trans. Spatial Algorithms Syst. **1**, article 3, 28 pages (2015)
3. M. Ahmed, S. Karagiorgou, D. Pfoser, C. Wenk, A comparison and evaluation of map construction algorithms using vehicle tracking data. GeoInformatica **19**(3), 601–632 (2015)
4. M. Ahmed, S. Karagiorgou, D. Pfoser, C. Wenk, *Map Construction Algorithms* (Springer, Berlin, 2015)
5. H. Alt, L.J. Guibas, Discrete geometric shapes: matching, interpolation, and approximation - a survey, in *Handbook of Computational Geometry*, ed. by J.-R. Sack, J. Urrutia (Elsevier, North-Holland, 1999), pp. 121–154
6. P. Bendich, E. Gasparovic, J. Harer, R. Izmailov, L. Ness, Multi-scale local shape analysis for feature selection in machine learning applications, in *Proceedings of International Joint Conference on Neural Networks* (2015)
7. J. Biagioni, J. Eriksson, Inferring road maps from global positioning system traces: survey and comparative evaluation. Transp. Res. Rec. J. Transp. Res. Board **2291**, 61–71 (2012)
8. G. Carlsson, Topology and data. Bull. Am. Math. Soc. **46**(2), 255–308 (2009)
9. F. Chazal, D. Cohen-Steiner, M. Glisse, L.J. Guibas, S.Y. Oudot, Proximity of persistence modules and their diagrams, in *Proceedings of 25th Annual Symposium on Computational Geometry* (2009), pp. 237–246
10. F. Chazal, V. De Silva, M. Glisse, S. Oudot, *The Structure and Stability of Persistence Modules* (Springer, Berlin, 2016)
11. O. Cheong, J. Gudmundsson, H.-S. Kim, D. Schymura, F. Stehn, Measuring the similarity of geometric graphs, in *Proceedings of International Symposium on Experimental Algorithms* (2009), pp. 101–112
12. S. Chowdhury, F. Mémoli, Persistent homology of directed networks, in *Proceedings of 50th Asilomar Conference on Signals, Systems and Computers* (IEEE, New York, 2016), pp. 77–81
13. D. Cohen-Steiner, H. Edelsbrunner, J. Harer, Stability of persistence diagrams, in *Proceedings of 21st Annual Symposium on Computational Geometry* (2005), pp. 263–271

14. D. Conte, P. Foggia, C. Sansone, M. Vento, Thirty years of graph matching in pattern recognition. Int. J. Pattern Recognit. Artif. Intell. **18**(3), 265–298 (2004)
15. H. Edelsbrunner, J. Harer, *Computational Topology: An Introduction* (AMS, Providence, 2010)
16. D. Eppstein, Subgraph isomorphism in planar graphs and related problems. J. Graph Algorithms Appl. **3**(3), 1–27 (1999)
17. R. Ghrist, Barcodes: the persistent topology of data. Bull. Am. Math. Soc. **45**, 61–75 (2008)
18. A. Grigor'yan, Y. Lin, Y. Muranov, S.-T. Yau, Homologies of path complexes and digraphs (2012, Preprint). arXiv:1207.2834
19. A. Hatcher, *Algebraic Topology* (Cambridge University Press, Cambridge, 2002). Electronic Version
20. S. Karagiorgou, D. Pfoser, On vehicle tracking data-based road network generation, in *Proceedings of 20th ACM SIGSPATIAL International Conference on Advances in Geographic Information Systems* (2012), pp. 89–98
21. H. Kashima, K. Tsuda, A. Inokuchi, Kernels for graphs, in *Kernel Methods in Computational Biology*, ed. by B. Schölkopf, K. Tsuda, J.-P. Vert (MIT Press, Cambridge, 2004), pp. 155–170
22. J. Mondzech, M. Sester, Quality analysis of Openstreetmap data based on application needs. Cartographica **46**, 115–125 (2011)
23. E. Munch, A user's guide to topological data analysis. J. Learn. Anal. **4**(2), 47–61 (2017)
24. J.R. Munkres, *Elements of Algebraic Topology* (Addison-Wesley, Redwood City, 1984)
25. Open street map. http://www.openstreetmap.org
26. R.C. Read, D.G. Corneil, The graph isomorphism disease. J. Graph Theory **1**(4), 339–363 (1977)
27. B. Schweinhart, J.K. Mason, R.D. MacPherson, Topological similarity of random cell complexes and applications. Phys. Rev. E **93**(6), 062111 (2016)
28. K. Turner, Generalizations of the Rips filtration for quasi-metric spaces with persistent homology stability results (2016, Preprint). arXiv:1608.00365
29. S.V.N. Vishwanathan, N.N. Schraudolph, R. Kondor, K.M. Borgwardt, Graph kernels. J. Mach. Learn. Res. **11**(Apr), 1201–1242 (2010)
30. Y. Xu, S.M. Salapaka, C.L. Beck, A distance metric between directed weighted graphs, in *2013 IEEE 52nd Annual Conference on Decision and Control* (IEEE, New York, 2013), pp. 6359–6364

Sweeping Costs of Planar Domains

Brooks Adams, Henry Adams, and Colin Roberts

Abstract Let D be a Jordan domain in the plane. We consider a pursuit-evasion, contamination clearing, or sensor sweep problem in which the pursuer at each point in time is modeled by a continuous curve, called the *sensor curve*. Both time and space are continuous, and the intruders are invisible to the pursuer. Given D, what is the shortest length of a sensor curve necessary to provide a sweep of domain D, so that no continuously-moving intruder in D can avoid being hit by the curve? We define this length to be the *sweeping cost* of D. We provide an analytic formula for the sweeping cost of any Jordan domain in terms of the geodesic Fréchet distance between two curves on the boundary of D with non-equal winding numbers. As a consequence, we show that the sweeping cost of any convex domain is equal to its width, and that a convex domain of unit area with maximal sweeping cost is the equilateral triangle.

1 Introduction

Let D be a Jordan domain, that is, the homeomorphic image of a disk in the plane. Suppose that continuously-moving intruders wander in D. You and a friend are each given one end of a rope, and your task is to drag this rope through the domain D in such a way so that every intruder is eventually intersected or caught by the rope. What is the shortest rope length you need in order to catch every possible intruder? We refer to such a continuous rope motion as a *sweep* of D (Fig. 1), and we refer to the length of the shortest such possible rope as the *sweeping cost* of D.

B. Adams · H. Adams (✉) · C. Roberts
Colorado State University, Department of Mathematics, Fort Collins, CO, USA
e-mail: brooksad@rams.colostate.edu; adams@math.colostate.edu

© The Author(s) and the Association for Women in Mathematics 2018
E. W. Chambers et al. (eds.), *Research in Computational Topology*, Association
for Women in Mathematics Series 13, https://doi.org/10.1007/978-3-319-89593-2_5

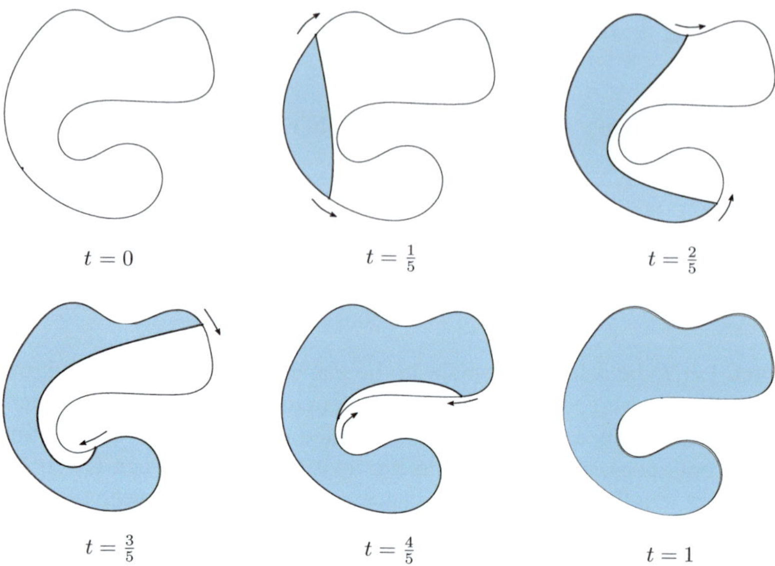

$t = 0$ $t = \frac{1}{5}$ $t = \frac{2}{5}$

$t = \frac{3}{5}$ $t = \frac{4}{5}$ $t = 1$

Fig. 1 An example sweep of a domain in the plane. A time $t = 0$ the entire domain is contaminated, and at time $t = 1$ the clearing sweep is complete

The problem we consider is only one example of a wide variety of interesting pursuit-evasion problems; see Sect. 2 for a brief introduction or [15], for example, for a survey. It is a pursuit-evasion problem in which both space and time are continuous, the pursuer is modeled at each point in time by a continuous curve, the intruder has complete information about the pursuer's location and its planned future movements, and the pursuer has no knowledge of the intruder's movements. Our problem can also be phrased as a contamination-clearing task, in which one must find the shortest rope necessary to clear domain D of a contaminant which, when otherwise unrestricted by the rope, moves at infinite speed to fill its region.

As first examples, the sweeping cost of a disk is equal to its diameter, and the sweeping cost of an ellipse is equal to the length of its minor axis. These computations follow from Theorem 5.4, in which we prove that the sweeping cost of domain D is at least as large as the shortest area-bisecting curve in D.

One motivation for considering a pursuer which is a rope, or a continuous curve at each point in time, is the context of mobile sensor networks. Suppose there is a large collection of disk-shaped sensors moving inside a planar domain, as considered in [1, 18]. What is the minimal number of sensors needed to clear this domain of all possible intruders? If n is the number of sensors, and $\frac{1}{n}$ is the diameter of each sensor, then as $n \to \infty$ an upper bound for the number of sensors needed is given by the sweeping cost (Fig. 2).

As our main result, in Theorem 7.1 we provide an analytic formula for the sweeping cost of an arbitrary Jordan domain D. Indeed, the sweeping cost of D is equal to the infimum, taken over all pairs of curves in the boundary of D whose

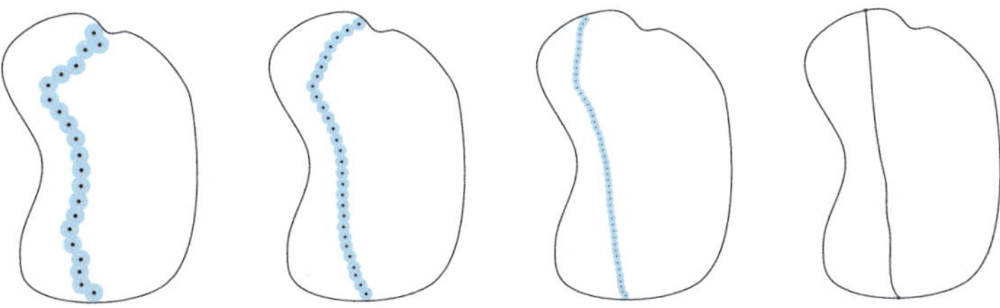

Fig. 2 If we have n disk-shaped sensors each of diameter $\frac{1}{n}$, then as $n \to \infty$ the sweeping cost gives an upper bound for the number of sensors needed to clear the domain of intruders

concatenation wraps a nontrivial number of times around the boundary, of the weak geodesic Fréchet distance between the two curves. The geodesic Fréchet distance differs from the standard Fréchet distance in that the distance between two points in D is not their Euclidean distance, but instead the length of the shortest path between them in D. Our Theorem 7.1 is also closely related to the geodesic width between two curves [19].

Using our main result, we prove in Theorem 8.1 that the sweeping cost of a convex domain D is equal to the width of D. As a consequence, it follows that the sweeping cost of a polygonal convex domain with n vertices can be computed in time $O(n)$ and space $O(n)$ using the *rotating calipers* technique [31]. Furthermore, it follows from [11, 27] that a convex domain of unit area with the maximal possible sweeping cost—that is, the most expensive convex domain to clear with a rope—is the equilateral triangle.

An intriguing open question motivated by our work is the following (Question 7.5). Given a Jordan domain D and two continuous injective curves α, β with image in the boundary ∂D, is the weak geodesic Fréchet distance between α and β equal to the strong geodesic Fréchet distance? The weak version of the Fréchet distance allows α and β to be reparameterized non-injectively, whereas the strong version does not.

We review related work in Sect. 2, state our problem of interest in Sect. 3, and describe some basic properties of the sweeping cost in Sect. 4. In Sect. 5, we provide a lower bound on the sweeping cost in terms of the shortest area-bisecting curves. We provide analytic formulas for the sweeping costs of Jordan and convex domains in Sects. 7 and 8, and in Sect. 9 we deduce that the sweeping cost of a unit-area convex domain is maximized by the equilateral triangle. The conclusion describes related problems of interest, and the appendix contains two technical lemmas and their proofs.

2 Related Work

A wide variety of pursuit-evasion problems have appeared in the mathematics, computer science, engineering, and robotics literature; see [15] for a survey. Space can modeled in a discrete fashion, for example by a graph [3, 7], or as a continuous domain in Euclidean space as we consider here. Time can similarly be discrete (turn-based) or continuous, as is our case. See [2, 6, 14, 17, 25, 28] for a selection of such problems.

A further important distinction in a pursuit-evasion problem is whether information is complete (pursuers and intruders know each other's locations), incomplete (pursuers and intruders are invisible to each other), or somewhere in-between. Our problem can be considered as one in which the pursuer has no knowledge of the intruders' movements, whereas the intruders have complete knowledge of the pursuer's current position and future movements. In other words, the pursuer must catch every possible intruder. Evasion problems in which the pursuer has no information and the intruders have complete information can equivalently be cast as contamination-clearing problems, see for example [1, 5, 18]. Indeed, the contaminated region of the domain at a particular time includes all locations where an intruder could currently be located, and the uncontaminated region is necessarily free of intruders. It is the task of the pursuer to clear the entire domain of contamination, so that no possible intruders could remain undetected.

The paper [19] introduces the *geodesic width* between two polylines, a notion that is very relevant for our problem. The geodesic width between two curves α, β is the same as the strong geodesic Fréchet distance between them (Definition 3.3) when domain D is chosen to be a region with a boundary consisting of curves α, β, and the two shortest paths connecting the endpoints of α and β. If curves α and β are polylines with n vertices in total, then [19] gives an $O(n^2 \log n)$ algorithm for computing the geodesic width between them. The goal of our paper is instead to rigorously prove an analytic formula for the sweeping cost of a domain, Theorem 7.1, that is closely related to geodesic widths. Indeed, the right-hand side of (6) in Theorem 7.1 is unchanged if we replace the geodesic distance between two curves (Definition 3.1) with the weak geodesic Fréchet distance between them (Definition 3.3). The paper [19] also studies sweeps of planar domains by piecewise linear curves in which the cost of a sweep is not equal to a length, but instead to the number of vertices or joints in the curve.

Related notions to the geodesic width include the isotopic Fréchet distance [13] and the minimum deformation area [12] between two curves α and β. Whereas the geodesic width considers deformations between α and β such that no two intermediate curves intersect, this restriction is not present for the isotopic Fréchet distance, which can therefore be defined between intersecting curves. The paper [12] considers a distance between two curves on a two-manifold which is instead an area: the minimal total surface area swept out by any deformation between the two curves. If the curves are piecewise linear in the plane, have n total vertices, and have I intersection points, then [12] gives an $O(n \log n + I^2 \log I)$ algorithm to compute the minimum deformation area between them. This work has been extended to different classes of curves by Fasy et al. [21] and Nie [26].

3 Preliminaries and Notation

Let $d\colon \mathbb{R}^2 \times \mathbb{R}^2 \to \mathbb{R}$ denote the Euclidean metric on $\mathbb{R}^2$. The distance between two subsets $X, Y \subseteq \mathbb{R}^2$ is defined as $d(X, Y) = \inf\{d(x, y) \mid x \in X \text{ and } y \in Y\}$. We denote the closure of a set $X \subseteq \mathbb{R}^2$ by $\overline{X}$.

3.1 Jordan Domains and Geodesics

Let $D \subseteq \mathbb{R}^2$ be a Jordan domain, that is, the homeomorphic image of a closed disk in $\mathbb{R}^2$. It follows that D is compact and simply-connected, and its boundary ∂D is a topological circle. Given a point $x \in D$, we let $B(x, \epsilon) = \{y \in D \mid d(x, y) < \epsilon\}$ denote the open ball about x in D. We denote the ϵ-offset of a set $X \subseteq D$ by $B(X, \epsilon) = \cup_{x \in X} B(x, \epsilon)$. Given a subset $X \subseteq D$, we define its boundary as $\partial X = \overline{X} \cap \overline{\mathbb{R}^2 \setminus X}$.

We refer the reader to [10] for the basics of geodesic curves and distances. The length of a continuous path $\gamma\colon [a, b] \to D$ is defined as in [10, Defintion 2.3.1]; we denote this length by $L(\gamma)$. Curve γ is said to be *rectifiable* if $L(\gamma) < \infty$. Domain D has a *length structure* [10, Section 2.1] in which all continuous paths are admissible, and the length is given by the function L. The associated *geodesic metric* $d_L\colon D \times D \to \mathbb{R}$, also known as a *path-length* or *intrinsic metric*, is

$$d_L(x, y) = \inf\{L(\gamma) \mid \gamma\colon [a, b] \to D \text{ is continuous with } \gamma(a) = x,\ \gamma(b) = y\}.$$

The precise definition of a geodesic, or length-minimizing curve in D, is given in [10, Definition 2.5.27]. Since D is a Jordan domain, it follows from [8, 9] that each pair of points in D is joined by a unique shortest geodesic in D.

Definition 3.1 Let $D \subseteq \mathbb{R}^2$ be a Jordan domain. We define the geodesic distance between two curves $\alpha, \beta\colon [t_0, t_1] \to D$ to be

$$d_L(\alpha, \beta) = \max_{t \in [t_0, t_1]} d_L(\alpha(t), \beta(t)).$$

3.2 Fréchet and Geodesic Fréchet Distances

The *Fréchet distance* is a measure of similarity between two curves $\alpha, \beta\colon [0, 1] \to \mathbb{R}^2$. One application of the Fréchet distance is in handwriting input recognition for a computer [30]: In order to properly tell which letters a user has written, the machine must determine which curves (representing letters) are the most similar. Other notions of distance, such as the Hausdorff distance between the images of the curves, are not necessarily sensitive enough for this task.

The intuition behind the Fréchet distance is that you are walking along path α, your dog is walking along path β, and you want to know how long of a leash you need. There are two notions, namely, the *weak Fréchet distance* and the *strong Fréchet distance*. In the weak case, you and your dog are allowed to backtrack along your respective paths, but in the strong case backtracking is forbidden. In general, these two distances need not be equal.

Definition 3.2 Let $\alpha, \beta \colon [0, 1] \to \mathbb{R}^2$ be continuous curves. Then the *weak (resp. strong) Fréchet distance* between α and β is

$$d_{\text{Fréchet}}(\alpha, \beta) = \inf_{a,b} \max_{t \in [0,1]} \{d\left(\alpha(a(t)), \beta(b(t))\right)\},$$

where the infimum is taken over all continuous $a, b \colon [0, 1] \to [0, 1]$ which are surjective (resp. bijective).

If $D \subseteq \mathbb{R}^2$ is a Jordan domain and $\alpha, \beta \colon [0, 1] \to D$ are two curves, then we can consider a variant of the Fréchet distance in which the Euclidean metric d is replaced with the geodesic metric d_L.

Definition 3.3 Let $D \subseteq \mathbb{R}^2$ be a Jordan domain, and let $\alpha, \beta \colon [0, 1] \to D$ be continuous curves. Then the *weak (resp. strong) geodesic Fréchet distance* between α and β is

$$d_{\text{geodesic Fréchet}}(\alpha, \beta) = \inf_{a,b} \max_{t \in [0,1]} \{d_L\left(\alpha(a(t)), \beta(b(t))\right)\},$$

where the infimum is taken over all continuous $a, b \colon [0, 1] \to [0, 1]$ which are surjective (resp. bijective).

See [32] for some algorithms related to computing geodesic Fréchet distances.

3.3 Sensor Curves

Let $I = [0, 1]$ be the unit interval. We define a sensor curve to be a time-varying rectifiable curve in D.

Definition 3.4 A *sensor curve* is a continuous map $f \colon I \times I \to D$ such that

(i) Each curve $f(\cdot, t) \colon I \to D$ is rectifiable and injective for $t \in (0, 1)$,
(ii) $f(I, 0)$ and $f(I, 1)$ are each (possibly distinct) single points in ∂D, and
(iii) $f(s, t) \in \partial D$ implies $s \in \{0, 1\}$ or $t \in \{0, 1\}$.

We think of the first input s as a spatial variable and of the second t as a temporal variable; in particular, $f(I, t)$ is the region covered by the curve of sensors at time t. Assumption (ii) states that the images of the sensor curve at times 0 and 1 are single points, and assumption (iii) implies that (apart from times 0 and 1) only the boundary of the sensor curve intersects ∂D. We define the length of a sensor curve to be

$$L(f) = \max_{t \in I} L(f(\cdot, t)).$$

An *intruder* is a continuous path $\gamma : I \to D$. We say that an intruder is caught by a sensor curve f at time t if $\gamma(t) \in f(I, t)$. A path $\gamma : [0, t] \to D$ such that $\gamma(t') \notin f(I, t')$ for all $t' \in [0, t]$ is called an *evasion path*. Sensor curve f is a *sweep* if every continuously moving intruder $\gamma : I \to D$ is necessarily caught at some time t, or equivalently, if no evasion path over the full time interval I exists.[1]

The following notation will prove convenient. Fix a sensor curve f. We let $C(t) \subseteq D$ be the *contaminated region* at time t, and we let $U(t) \subseteq D$ be the *uncontaminated region* at time t. More precisely,

$$C(t) = \left\{ \begin{array}{c} x \in D \mid \exists\, \gamma : [0, t] \to D \text{ with } \gamma(t) = x \\ \text{and } \gamma(t') \notin f(I, t')\ \forall t' \in [0, t] \end{array} \right\} \quad \text{and} \quad U(t) = D \setminus C(t).$$

Note that sensor curve f is a sweep if and only if $C(1) = \emptyset$, or equivalently $U(1) = D$.

Definition 3.5 Let $\mathcal{F}(D)$ be the set of all sensor curve sweeps of D. The *sweeping cost* of D is

$$SC(D) = \inf_{f \in \mathcal{F}(D)} L(f).$$

Remark 3.6 The results of Sects. 4 and 5 hold even if assumptions (ii) and (iii) in Definition 3.4 are removed.

4   Properties of Sensor Sweeps

We now prove some basic properties of sensor sweeps and the contaminated and uncontaminated regions.

Lemma 4.1 *If x and x' are in the same path-connected component of $D \setminus f(I, t)$, then $x \in U(t)$ if and only if $x' \in U(t)$.*

Proof Suppose for a contradiction that $x \in U(t)$ but $x' \notin U(t)$. Since $x' \in C(t)$, there exists an evasion path $\gamma : [0, t] \to D$ with $\gamma(t) = x'$. Since x and x' are in the same path-connected component, there exists a path $\beta : I \to D \setminus f(I, t)$ with $\beta(0) = x$ and $\beta(1) = x'$.

---

[1] Our definition is similar to the graph-based definition in [3, Definition 2.1].

Note that $\beta(I)$ and $f(I, t)$ are compact, since they are each a continuous image of the compact set I. Since D is a metric space it is normal, and therefore there exist disjoint neighborhoods containing $\beta(I)$ and $f(I, t)$. Because f is uniformly continuous (it is a continuous function on a compact set), we can choose $\delta_1 > 0$ such that $f(I, [t - \delta_1, t])$ remains in this open neighborhood disjoint from $\beta(I)$, giving

$$\beta(I) \cap f(I, [t - \delta_1, t]) = \emptyset. \tag{1}$$

Again by normality of D, there exist disjoint neighborhoods containing $\gamma(t) = x'$ and $f(I, t)$. Since γ is continuous and f is uniformly continuous, we can choose $\delta_2 > 0$ such that $\gamma([t - \delta_2, t])$ and $f(I, [t - \delta_2, t])$ remain in these disjoint neighborhoods, giving

$$\gamma([t - \delta_2, t]) \cap f(I, [t - \delta_2, t]) = \emptyset. \tag{2}$$

Let $\delta = \min\{\delta_1, \delta_2\}$. Using (1) and (2) we can define an evasion path $\gamma : [0, t] \to D$ with $\gamma(t) = x$. Indeed, let

$$\gamma(t') = \begin{cases} \gamma(t') & \text{if } t' \leq t - \delta \\ \gamma(2t' - t + \delta) & \text{if } t - \delta < t' \leq t - \frac{\delta}{2} \\ \beta(\frac{2}{\delta}(t' - t + \frac{\delta}{2})) & \text{if } t - \frac{\delta}{2} < t' \leq t. \end{cases}$$

This contradicts the fact that $x \in U(t)$.   □

Lemma 4.2 *For all $t \in I$, the set $U(t)$ is closed and the set $C(t)$ is open in D.*

Proof Suppose $x \in C(t)$. Since $f(I, t)$ is closed and $x \notin f(I, t)$, there exists some $\varepsilon > 0$ such that $B(x, \varepsilon) \cap f(I, t) = \emptyset$. Note all $x' \in B(x, \varepsilon)$ are in the same path-connected component of $D \setminus f(I, t)$ as x via a straight line path. Hence Lemma 4.1 implies $B(x, \varepsilon) \subseteq C(t)$, showing $C(t)$ is open in D. It follows that $U(t) = D \setminus C(t)$ is closed in D.   □

Lemma 4.3 *If $h : D \to h(D)$ is a homeomorphism onto its image $h(D) \subseteq \mathbb{R}^2$, then a sensor curve $f : I \times I \to D$ is a sweep of D if and only if sensor curve $hf : I \times I \to h(D)$ is a sweep of $h(D)$.*

Proof (See Fig. 3). Note that if $\gamma : I \to D$ is an evasion path for f, then $h\gamma : I \to h(D)$ is an evasion path for hf. Conversely, if $\gamma : I \to h(D)$ is an evasion path for hf, then $h^{-1}\gamma : I \to D$ is an evasion path for f.   □

Fig. 3 A homeomorphism
$h\colon D \to h(D)$

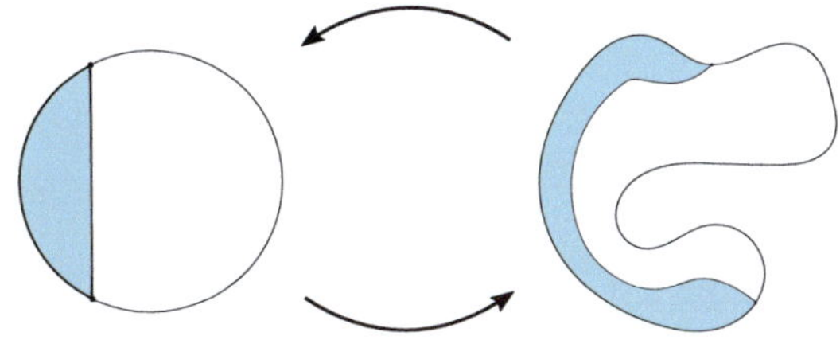

5   A Lower Bound on the Sweeping Cost

In this section, we prove that the sweeping cost of a Jordan domain is at least as large as the length of the shortest area-bisecting curve. The first lemma is a version of the intermediate value theorem with slightly relaxed hypotheses.

Lemma 5.1 *If $f\colon [a, b] \to \mathbb{R}$ is upper semicontinuous and left continuous and if $f(a) < u < f(b)$, then there exists some $c \in (a, b)$ with $f(c) = u$.*

Proof Let S be the set of all $x \in (a, b)$ with $f(x) < u$. Then S is non-empty since $a \in S$, and S is bounded above by b. Hence by the completeness of $\mathbb{R}$, the supremum $c = \sup S$ exists. We claim that $f(c) = u$.

Let $\epsilon > 0$. Since f is left-continuous, there is some $\delta > 0$ such that $|f(x) - f(c)| < \epsilon$ whenever $x \in (c - \delta, c]$. By the definition of supremum, there exists some $y \in (c - \delta, c]$ that is contained in S, giving $f(c) < f(y) + \epsilon < u + \epsilon$. Since this is true for all $\epsilon > 0$, it follows that $f(c) \le u$.

It remains to show $f(c) \ge u$. Let $\epsilon > 0$. Since f is upper semicontinuous, there exists a $\delta > 0$ such that $f(c) > f(x) - \epsilon$ whenever $x \in (c - \delta, c + \delta)$. Let $y \in (c, c + \delta)$ and note that $y \notin S$, giving $f(c) > f(y) - \epsilon \ge u - \epsilon$. It follows that $f(c) \ge f(u)$. □

If $S \subseteq \mathbb{R}^2$ is a measurable set, then we let area(S) denote its area.

Lemma 5.2 *The function* area($U(t)$) *is upper semicontinuous and left continuous.*

Proof Let $t_0 \in I$. We will show that area($U(t)$) is right upper semicontinuous and left continuous at t_0, which implies the function is both upper semicontinuous and left continuous.

For right upper semicontinuity, note for $t \ge t_0$, we have

$$U(t) \subseteq U(t_0) \cup f(I, [t_0, t]). \tag{3}$$

Since sensor curve $f\colon I \times I \to D$ is a continuous function on a compact domain, it is also uniformly continuous. Hence for all $\epsilon > 0$ there exists some δ such that

$$f(I, [t_0, t_0 + \delta]) \subseteq B(f(I, t_0), \epsilon). \tag{4}$$

It follows that for all $t \in [t_0, t_0 + \delta]$, we have

$$\text{area}(U(t)) - \text{area}(U(t_0)) \le \text{area}(f(I, [t_0, t])) \qquad \text{by (3)}$$

$$\le \text{area}(B(f(I, t_0), \epsilon)) \qquad \text{by (4)}$$

$$\le 2L(f(I, t_0))\epsilon + \pi\epsilon^2,$$

where the last inequality is by a result of Hotelling (see e.g., [23, Equation (2.1)]). Hence area(t) is right upper semicontinuous.

To see that area($U(t)$) is left continuous at $t_0 \in I$, we must show that for all sequences $\{s_i\}$ with $0 \le s_i \le t_0$ and $\lim_i s_i = t_0$, we have $\lim_i \text{area}(U(s_i)) = \text{area}(U(t_0))$. We claim

$$U(t_0) \setminus f(I, t_0) \subseteq \liminf_i U(s_i) \subseteq \limsup_i U(s_i) \subseteq U(t_0), \qquad (5)$$

where the middle containment is by definition. We now justify the first and last containment.

To prove $U(t_0) \setminus f(I, t_0) \subseteq \liminf_i U(s_i)$ it suffices to show that for any $x \in U(t_0) \setminus f(I, t_0)$ there exists an $\epsilon > 0$ such that $x \in U(t_0 - \delta)$ for all $\delta \in [0, \epsilon)$. Fix ϵ such that $x \notin f(I, t)$ for $t \in (t_0 - \epsilon, t_0]$. Suppose for a contradiction that $x \in C(t_0 - \delta)$ for some $\delta \in [0, \epsilon)$. Hence there exists an evasion path $\gamma : [0, t_0 - \delta] \to D$ with $\gamma(t_0 - \delta) = x$. It is possible to extend γ to an evasion path $\tilde{\gamma} : [0, t_0] \to D$ defined by

$$\tilde{\gamma}(t) = \begin{cases} \gamma(t) & \text{if } t \in [0, t_0 - \delta] \\ x & \text{if } t \in (t_0 - \delta, t_0]. \end{cases}$$

This contradicts the fact $x \in U(t_0)$, thus giving the first containment.

We now show $\limsup_i U(s_i) \subseteq U(t_0)$. If $x \notin U(t_0)$, then there exists an evasion path $\gamma : [0, t_0] \to D$ with $\gamma(t_0) = x$. Since $C(t_0)$ is open by Lemma 4.2, there exists some $\delta > 0$ such that $B(x, \delta) \subseteq C(t_0)$, and hence $B(x, \delta) \cap f(I, t_0) = \emptyset$. Since f is uniformly continuous, there is some $\epsilon_1 > 0$ sufficiently small with $B(x, \delta) \cap f(I, [t_0 - \epsilon_1, t_0]) = \emptyset$, and since γ is continuous there is some $\epsilon_2 > 0$ with $\gamma([t_0 - \epsilon_2, t_0]) \subseteq B(x, \delta)$. Let $\epsilon = \min\{\epsilon_1, \epsilon_2\}$. Reparameterize γ to get a continuous curve $\tilde{\gamma} : [0, t_0] \to D$ with

$$\tilde{\gamma}(t) = \begin{cases} \gamma(t) & \text{if } x \in [0, t_0 - \epsilon) \\ \gamma(t) \in B(x, \delta) & \text{if } t \in [t_0 - \epsilon, t_0 - \epsilon/2) \\ \gamma(t) = x & \text{if } t \in [t_0 - \epsilon/2, t_0]. \end{cases}$$

The evasion path $\tilde{\gamma}$ shows $x \notin \limsup_i U(s_i)$, giving the third containment and finishing the proof of (5).

Set $f(I, t_0)$ has Lebesgue measure zero since curve $f(\cdot, t_0)$ is rectifiable, giving area($U(t_0) \setminus f(I, t_0)$) = area($U(t_0)$). Thus (5) implies

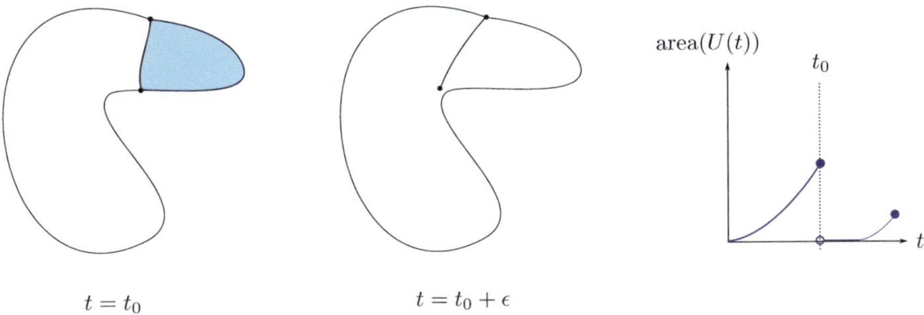

Fig. 4 An example sensor curve where the function area$(U(t))$ is not right continuous. The shaded region is $U(t)$ and the unshaded region is $C(t)$

$$\text{area}(\limsup_i U(s_i)) = \text{area}(U(t_0)) = \text{area}(\liminf_i U(s_i)).$$

Since area(D) is finite, Lemma 1 in the appendix implies $\limsup_i \text{area}(U(s_i)) \leq \text{area}(\limsup_i U(s_i))$ and $\text{area}(\liminf_i U(s_i)) \leq \liminf_i \text{area}(U(s_i))$, giving

$$\limsup_i \text{area}(U(s_i)) \leq \text{area}(U(t_0)) \leq \liminf_i \text{area}(U(s_i)).$$

Hence $\lim_i \text{area}(U(s_i)) = \text{area}(U(t_0))$ as required. $\square$

Remark 5.3 The function area$(U(t))$ need not be right continuous. Indeed, consider a sensor curve as shown below in Fig. 4, where area$(U(t_0)) > 0$, and where there is some $\epsilon_0 > 0$ such that for all $0 < \epsilon < \epsilon_0$, only one point on the sensor curve at time $t_0 + \epsilon$ intersects ∂D and area$(U(t_0 + \epsilon)) = 0$.

As a consequence, we obtain the following lower bound on the sweeping cost.

Theorem 5.4 *If D is a Jordan domain, then the sweeping cost* SC(D) *is at least as large as the length of the shortest area-bisecting curve in D.*

Proof Suppose that f is a sweep of D. Note that area$(U(0)) = 0$ and area$(U(1)) = $ area(D). By Lemmas 5.1 and 5.2, there exists some time $t' \in I$ with area$(U(t')) = \frac{1}{2}$area(D). So $L(f(\cdot, t'))$ and hence SC(D) is at least as large as the shortest area-bisecting curve in D. $\square$

Example 5.5 If $D = \{(x, y) \in \mathbb{R}^2 \mid x^2 + y^2 \leq 1\}$ is the unit disk, then SC$(D) = 2$.

Proof To see SC$(D) \leq 2$, consider the sweep $f : [-1, 1] \times I \rightarrow D$ defined by $f(s, t) = (2s\sqrt{t - t^2}, 2t - 1)$ (Fig. 5) which has length 2.

For the reverse direction, note that the shortest area-bisecting curve in D is a diameter [20]. Hence we apply Theorem 5.4 to get SC$(D) \geq 2$. $\square$

Example 5.6 Let $a, b > 0$. If $D = \{(x, y) \in \mathbb{R}^2 \mid (x/a)^2 + (y/b)^2 \leq 1\}$ is the convex hull of an ellipse, then SC$(D) = \min\{2a, 2b\}$.

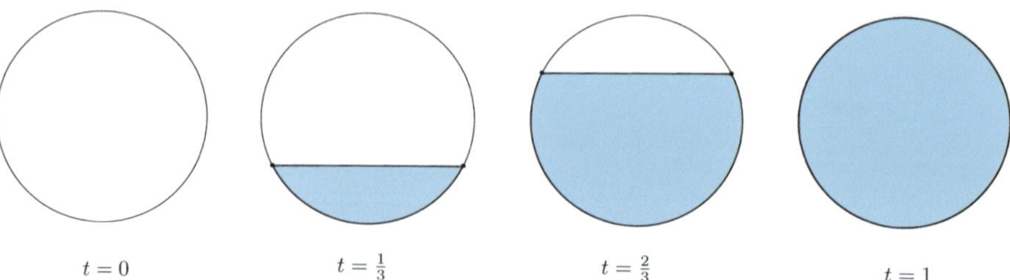

Fig. 5 A sweep of the unit disk. The shaded region is $U(t)$ and the unshaded region is $C(t)$

Proof To see $\mathrm{SC}(D) \leq \min\{2a, 2b\}$, construct a sweep much like in Example 5.5.

For the reverse direction, the solution to [29, Chapter X, Problem 33] states that because the ellipse has a center of symmetry, the shortest area-bisecting curve is a straight line. All area-bisecting lines pass through the center of the ellipse, and hence have length at least $\min\{2a, 2b\}$. It follows from Theorem 5.4 that $\mathrm{SC}(D) \geq \min\{2a, 2b\}$. $\qquad\square$

6 A Lemma of No Progress

In Sects. 7–9 we will restrict attention to sensor curves f with boundary points $f(0, t), f(1, t) \in \partial D$ for all $t \in I$. The motivation behind this assumption is Lemma 6.2, which states that if $f(0, t) \notin \partial D$ or $f(1, t) \notin \partial D$, then the uncontaminated region at time t is as small as possible, namely $U(t) = f(I, t)$.

The following lemma is from [33]; see also its statement in [24, p. 164].

Lemma 6.1 (Zoretti) *If K is a bounded maximal connected subset of a plane closed set M and $\epsilon > 0$, then there exists a simple closed curve J enclosing K such that $J \cap M = \emptyset$ and $J \subseteq B(K, \epsilon)$.*

Lemma 6.2 *Let $f : I \times I \to D$ be a sensor curve. If $f(0, t) \notin \partial D$ or $f(1, t) \notin \partial D$ and $U(t) \neq D$, then $U(t) = f(I, t)$.*

Proof Without loss of generality suppose $f(0, t) \notin \partial D$. It suffices to show that $D \setminus f(I, t)$ is a single path-connected component, because then Lemma 4.1 and the fact that $U(t) \neq D$ will imply $C(t) = D \setminus f(I, t)$ and hence $U(t) = f(I, t)$. Let $x, x' \in D \setminus f(I, t)$; we must find a path in $D \setminus f(I, t)$ connecting x and x'. There are two cases: when $f(1, t) \notin \partial D$, and when $f(1, t) \in \partial D$.

In the first case $f(1, t) \notin \partial D$, note that $f(I, t)$ is disjoint from ∂D. Hence by compactness there exists some $\epsilon > 0$ such that $d(f(I, t), \partial D \cup \{x, x'\}) < \epsilon$. By Lemma 6.1 (with $K = f(I, t)$ and $M = \partial D \cup \{x, x'\}$), there exists a simple closed curve J in D enclosing $f(I, t)$ but not enclosing x or x'. We may therefore connect x and x' by a path in $D \setminus f(I, t)$ consisting of three pieces: a path in D from x to J, a path in D from x' to J, and a path in J connecting these two endpoints (Fig. 6).

In the second case $f(1, t) \in \partial D$, pick some point $y \in D \setminus (f(I, t) \cup \{x, x'\})$. By translating D in the plane, we may assume that $y = \vec{0}$. Define the inversion function $i : \mathbb{R}^2 \setminus \{\vec{0}\} \to \mathbb{R}^2 \setminus \{\vec{0}\}$ by $i(r \cos\theta, r \sin\theta) = (\frac{1}{r} \cos\theta, \frac{1}{r} \sin\theta)$; note i^2 is the identity map. Let $\epsilon > 0$ be such that

$$d(i(f(I, t) \cup \partial D), \{i(x), i(x')\}) < \epsilon.$$

By Lemma 6.1 (with $K = i(f(I, t) \cup \partial D)$ and $M = \{i(x), i(x')\}$), there exists a simple closed curve J in $\mathbb{R}^2$ enclosing $i(f(I, t) \cup \partial D)$ but not enclosing $i(x)$ or $i(x')$. By the Jordan curve theorem, $i(x)$ and $i(x')$ are in the same (exterior) connected component E of $\mathbb{R}^2 \setminus J$. Since E is open it is also path-connected, and hence we can connect $i(x)$ and $i(x')$ by a path γ in E. The path $i(\gamma)$ is therefore a path in $D \setminus f(I, t)$ connecting x and x' (Fig. 7).

$\square$

7 Sweeping Cost of a Jordan Domain

As motivated by Lemma 6.2, for the remainder of the paper we restrict attention to sensor curves satisfying $f(s, t) \in \partial D$ if and only if $s \in \{0, 1\}$ or $t \in \{0, 1\}$.

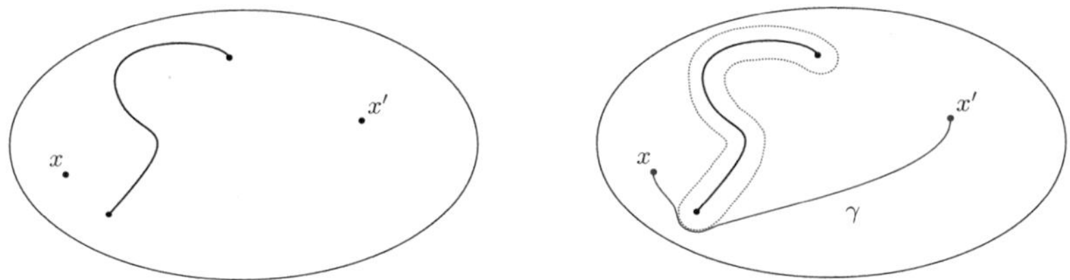

Fig. 6 The first case in the proof of Lemma 6.2, with J drawn in red

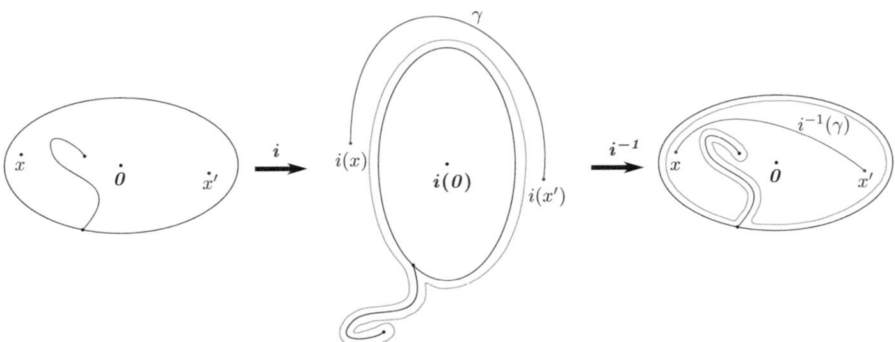

Fig. 7 The second case in the proof of Lemma 6.2, with J and $i(J)$ drawn in red

Given curves $\alpha, \beta \colon I \to \partial D$ with $\alpha(1) = \beta(0)$, we define the concatenated curve $\alpha \cdot \beta \colon I \to \partial D$ by

$$\alpha \cdot \beta(t) = \begin{cases} \alpha(2t) & \text{if } 0 \leq t < \frac{1}{2} \\ \beta(2t-1) & \text{if } \frac{1}{2} \leq t \leq 1. \end{cases}$$

We define the inverse curve $\beta^{-1} \colon I \to \partial D$ by $\beta^{-1}(t) = \beta(1-t)$.

Since ∂D is homeomorphic to the circle, given a loop $\gamma \colon I \to \partial D$ (with $\gamma(0) = \gamma(1)$) we can denote the winding number of γ, that is, the number of times γ wraps around ∂D, by $\mathrm{wn}(\gamma)$. The winding number is positive (resp. negative) for loops that wrap around in the counterclockwise (resp. clockwise) direction. Note that if α and β are paths in ∂D with $\alpha(0) = \beta(0)$ and $\alpha(1) = \beta(1)$, then $\alpha \cdot \beta^{-1}$ is a loop (see Fig. 8).

Our main result is an analytic formula for the sweeping cost of a Jordan domain.

Theorem 7.1 *The sweeping cost of a Jordan domain D is*

$$\mathrm{SC}(D) = \inf\{d_L(\alpha, \beta) \mid \alpha, \beta \colon I \to \partial D,\ \alpha(0) = \beta(0),\ \alpha(1) = \beta(1),\ \mathrm{wn}(\alpha \cdot \beta^{-1}) \neq 0\}. \tag{6}$$

Equation (6) is closely related to the *geodesic width* between two polylines [19], and also the *isotopic Fréchet distance* between two curves [13]. Indeed, note that the right-hand side of (6) is unchanged if we replace $d_L(\alpha, \beta)$ with the weak geodesic Fréchet distance between α and β (see Definition 3.3).

Remark 7.2 The value of the right-hand side of (6) is unchanged if we replace $\mathrm{wn}(\alpha \cdot \beta^{-1}) \neq 0$ with $\mathrm{wn}(\alpha \cdot \beta^{-1}) \in \{-1, 1\}$.

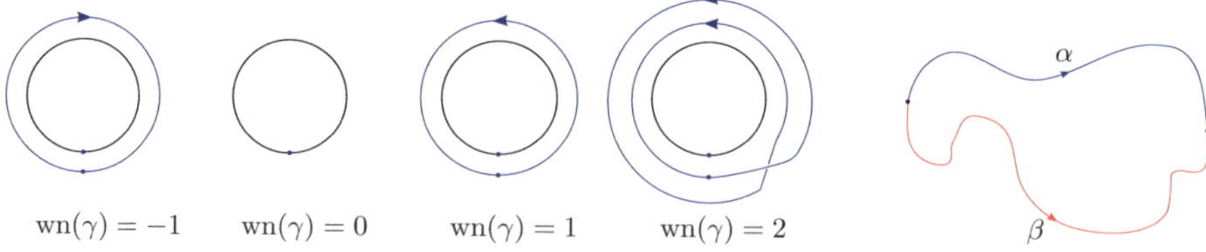

Fig. 8 (Left) Example winding numbers. (Right) Two curves $\alpha, \beta \colon I \to \partial D$ with $\alpha(0) = \beta(0)$ and $\alpha(1) = \beta(1)$; note $\alpha \cdot \beta^{-1}$ is a loop

Proof of Remark 7.2 Let $\alpha, \beta \colon I \to \partial D$ with $\alpha(0) = \beta(0)$ and $\alpha(1) = \beta(1)$, and suppose $|\mathrm{wn}(\alpha \cdot \beta^{-1})| \geq 2$. Hence there exists some $0 < t < 1$ such that $\alpha(t) = \beta(t)$ and $\mathrm{wn}(\alpha|_{[0,t]} \cdot \beta|_{[0,t]}^{-1}) \in \{-1, 1\}$. The claim follows since

$$d_L(\alpha|_{[0,t]}, \beta|_{[0,t]}) \leq d_L(\alpha, \beta).$$

□

The following lemma will be used to prove the $\leq$ direction in (6).

Lemma 7.3 *Let D be a Jordan domain. Suppose $\alpha, \beta \colon I \to \partial D$ with $\alpha(0) = \beta(0)$, $\alpha(1) = \beta(1)$, and $\mathrm{wn}(\alpha \cdot \beta^{-1}) \neq 0$. If $f \colon I \times I \to D$ is any sensor curve with $f(0, t) = \alpha(t)$ and $f(1, t) = \beta(t)$, then f is a sweep of D.*

Proof Let $\mathbb{D} = \{(x, y) \in \mathbb{R}^2 \mid x^2 + y^2 \leq 1\}$ be the unit disk. We first prove this claim in the case when $D = \mathbb{D}$.

By Lemma 2, in the appendix, there exists a point $p \in \mathbb{R}^2 \setminus \mathbb{D}$ and two continuous families of curves $g_\alpha, g_\beta \colon I \times I \to \mathbb{R}^2$ such that

- $g_\alpha(0, t) = p = g_\beta(0, t)$,
- $g_\alpha(1, t) = \alpha(t)$,
- $g_\beta(1, t) = \beta(t)$, and
- $g_\alpha(s, t), g_\beta(s, t) \notin \mathbb{D}$ for $s < 1$.

Let S^1 be the circle of unit circumference, that is, $[0, 1]$ with endpoints 0 and 1 identified. Define a continuous map $g \colon S^1 \times I \to D$ via

$$g(s, t) = \begin{cases} g_\alpha(3st, t) & \text{if } 0 \leq s < \frac{1}{3} \\ f(3s - 1, t) & \text{if } \frac{1}{3} \leq s < \frac{2}{3} \\ g_\beta(3t(1 - s), t) & \text{if } \frac{2}{3} \leq s \leq 1. \end{cases}$$

Note $g(\cdot, t)$ is indeed a (possibly non-simple) map from the circle since $g_\alpha(0, t) = p = g_\beta(0, t)$ for all t. Define a continuous signed distance $d^\pm \colon \mathbb{R}^2 \times I \to \mathbb{R}$ by

$$d^\pm(x, t) = \begin{cases} d(x, g(S^1, t)) & \text{if } x \in g(S^1, t) \text{ or } \mathrm{wn}(g(\cdot, t), x) = 0 \\ -d(x, g(S^1, t)) & \text{if } x \notin g(S^1, t) \text{ and } \mathrm{wn}(g(\cdot, t), x) \neq 0. \end{cases}$$

Here $\mathrm{wn}(g(\cdot, t), x)$ denotes (for $x \notin g(S^1, t)$) the winding number of the map $g(\cdot, t) \colon S^1 \to \mathbb{R}^2 \setminus \{x\} \simeq S^1$. Note that $\mathrm{wn}(g(\cdot, t), x)$ is constant on each connected component of $\mathbb{R}^2 \setminus g(S^1, t)$, and that $d^\pm$ is continuous.

Given any intruder path $\gamma \colon I \to \mathbb{D}$, the continuous function $d^\pm(\gamma(t), t) \colon I \to \mathbb{R}$ satisfies $d^\pm(\gamma(0), 0) \geq 0$ (since $f(I, 0)$ is a single point in ∂D) and $d^\pm(\gamma(1), 1) \leq 0$ (since $f(I, 1)$ is a single point in ∂D and $\mathrm{wn}(\alpha \cdot \beta^{-1}) \neq 0$). By the intermediate value theorem there exists some $t' \in I$ with $d^\pm(\gamma(t'), t') = 0$, and hence $\gamma(t') \in g(S^1, t') \cap \mathbb{D} = f(I, t')$. So γ is not an evasion path, and f is a sweep of $\mathbb{D}$.

We now handle the case when D is an arbitrary Jordan domain. By definition there exists a homeomorphism $h\colon \mathbb{D} \to D$. Note that $h^{-1}\alpha, h^{-1}\beta\colon I \to \partial\mathbb{D}$ with $h^{-1}\alpha(0) = h^{-1}\beta(0)$, $h^{-1}\alpha(1) = h^{-1}\beta(1)$, and $\mathrm{wn}(h^{-1}\alpha \cdot h^{-1}\beta^{-1}) \neq 0$. Since $h^{-1}f\colon I \times I \to \mathbb{D}$ is a sensor curve, it follows from our proof in the case of the disk that $h^{-1}f$ is a sweep of $\mathbb{D}$. Hence f is a sweep of D by Lemma 4.3. $\square$

Proof of Theorem 7.1 Let

$$c = \inf\{d_L(\alpha, \beta) \mid \alpha, \beta\colon I \to \partial D,\ \alpha(0) = \beta(0),\ \alpha(1) = \beta(1),\ \mathrm{wn}(\alpha \cdot \beta^{-1}) \neq 0\}.$$

We first prove the $\leq$ direction of (6). Let $\epsilon > 0$ be arbitrary. By the definition of infimum there exist curves $\alpha, \beta\colon I \to \partial D$ with $\alpha(0) = \beta(0)$, $\alpha(1) = \beta(1)$, $\mathrm{wn}(\alpha \cdot \beta^{-1}) \neq 0$, and $d_L(\alpha, \beta) \leq c + \epsilon$. Define the continuous map $f\colon I \times I \to D$ by letting $f(\cdot, t)\colon I \to D$ be the unique constant-speed geodesic in D between $f(0, t) = \alpha(t)$ and $f(1, t) = \beta(t)$, which exists by Bourgin and Renz [9], Bishop [8]. Lemma 7.3 implies that f is a sweep, and hence we have

$$\mathrm{SC}(D) \leq L(f) = d_L(\alpha, \beta) \leq c + \epsilon.$$

Since this is true for all $\epsilon > 0$, we have $\mathrm{SC}(D) \leq c$.

For the $\geq$ direction of (6), suppose that f is a sensor curve with $L(f) < c$. For notational convenience, define $\alpha, \beta\colon I \to \partial D$ by $\alpha(t) = f(0, t)$ and $\beta(t) = f(1, t)$. Then necessarily $\mathrm{wn}(\alpha \cdot \beta^{-1}) = 0$, and furthermore

$$\alpha(t) = \beta(t) \quad \text{implies} \quad \mathrm{wn}(\alpha|_{[0,t]} \cdot \beta|_{[0,t]}^{-1}) = 0, \tag{7}$$

since otherwise we'd have

$$L(f) \geq d_L(\alpha, \beta) \geq d_L(\alpha|_{[0,t]}, \beta|_{[0,t]}) \geq c,$$

a contradiction. We will show that f is not a sweep of D by showing the existence of an evasion path $\gamma\colon I \to D$ whose image furthermore lives in ∂D.

Indeed, consider the one-dimensional evasion problem in ∂D where the region covered by the sensors at time t is $\{\alpha(t), \beta(t)\}$. In this one-dimensional problem, it is clear that the uncontaminated region in ∂D is either (i) a single point $\alpha(t) = \beta(t)$, (ii) a closed interval in ∂D with endpoints $\alpha(t)$ and $\beta(t)$, or (iii) all of ∂D. Equation (7), however, rules out the possibility of (iii). It follows that the contaminated region in ∂D is always a non-empty open interval in ∂D with continuously varying endpoints $\alpha(t)$ and $\beta(t)$. Therefore we can define an evasion path $\gamma\colon I \to \partial D$, for example by letting $\gamma(t)$ be the midpoint of the open interval of the uncontaminated region in ∂D. This evasion path γ is also an evasion path for our original two-dimensional problem in D, as $\gamma\colon I \to \partial D \subseteq D$ satisfies $\gamma(t) \notin f(I, t)$ for all t. This gives the $\geq$ direction of (6). $\square$

Question 7.4 Does Theorem 7.1 hold even if assumption (iii) in Definition 3.4 is removed, that is, if the interior of a sensor curve is also allowed to touch ∂D?

Question 7.5 For any Jordan domain D in the plane and injective curves $\alpha, \beta \colon I \to \partial D$, we conjecture that the weak geodesic Fréchet distance between α and β is equal to their strong geodesic Fréchet geodesic distance.

There are simple counterexamples to Question 7.5 when α and β are not injective, or when they do not map to ∂D. Closely related is following question: Is the value of (6) unchanged if we require α and β to be injective?

8 Sweeping Cost of a Convex Domain

Given a convex Jordan domain $D \subseteq \mathbb{R}^2$, its *width* $w(D)$ is defined as

$$w(D) = \min_{\|v\|=1} \ \max_{x \in \mathbb{R}^2} \ L(D \cap \{x + tv \mid t \in \mathbb{R}\}),$$

where v is a unit direction vector in $\mathbb{R}^2$, and where $\{x + tv \mid t \in \mathbb{R}\}$ is a line through x in direction v. Alternatively, the width $w(D)$ is the smallest distance between two parallel supporting lines on opposite sides of D (Fig. 9).

Theorem 8.1 *If D is a convex Jordan domain, then* $\mathrm{SC}(D) = w(D)$.

Proof We first show $\mathrm{SC}(D) \leq w(D)$. By Theorem 7.1, it suffices to show

$$\inf\{d_L(\alpha, \beta) \mid \alpha, \beta \colon I \to \partial D, \ \alpha(0) = \beta(0), \alpha(1){=}\beta(1), \ \mathrm{wn}(\alpha{\cdot}\beta^{-1}) \neq 0\} \leq w(D).$$

Let v some direction vector realizing the width, that is, $w(D) = \max_{x \in \mathbb{R}^2} L(D \cap \{x + tv \mid t \in \mathbb{R}\})$. Consider sweeping through all lines in $\mathbb{R}^2$ parallel to v; the intersection of these lines with ∂D traces out two continuous curves $\alpha, \beta \colon I \to \partial D$

Fig. 9 The width $w(D)$ of a domain D

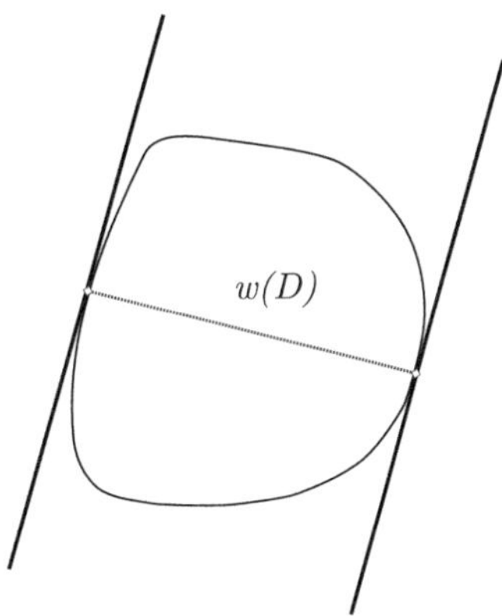

with $\alpha(0) = \beta(0)$, $\alpha(1) = \beta(1)$, and $\mathrm{wn}(\alpha \cdot \beta^{-1}) = \pm 1$.[2] We have $d_L(\alpha, \beta) \leq w(D)$, giving $\mathrm{SC}(D) \leq w(D)$.

To finish the proof, we need some background on planar convex domains. A point $x \in \partial D$ is *smooth* if it has a unique supporting hyperplane, and otherwise x is a *vertex* containing a range of angles $[\theta_1, \theta_2] \subseteq S^1$ (with $\theta_1 \neq \theta_2$) which are the outward normal directions of supporting hyperplanes of D at x. Away from the vertices, the unique supporting hyperplane of $x \in \partial D$ varies continuously with x. By Berger [4, Proposition 11.6.2], the set of vertices of the closed convex domain D is countable.

We now show $\mathrm{SC}(D) \geq w(D)$. Given $\epsilon > 0$, let α and β be ϵ-close to realizing the infimum in (6), meaning $\mathrm{SC}(D) + \epsilon \geq d_L(\alpha, \beta)$. For notational convenience, we assume that $\alpha(0) = \beta(0)$ and $\alpha(1) = \beta(1)$ are not vertices of ∂D (our same proof technique works regardless). Let $T = \{t_1, t_2, t_3, \ldots\} \subseteq I$ be a countable subset such that $t \in T$ if either $\alpha(t)$ or $\beta(t)$ is a vertex of ∂D. Let $t_0 = 0$, and if $|T|$ is finite, then let $t_{|T|+1} = 1$. For $i = 1, 2, \ldots, |T|$, choose weights $w_i > 0$ such that $\sum_i w_i = w < \infty$; this is possible since T is countable. Let $p_1: \partial D \times S^1 \to \partial D$ and $p_2: \partial D \times S^1 \to S^1$ be the projection maps. It is possible to define a *continuous* map $g_\alpha: [0, 1 + w] \to \partial D \times S^1$ satisfying the following properties.

- Each $g_\alpha(s)$ is equal to a point $(\alpha(t), v) \in \partial D \times S^1$ with $t \in I$ such that v is the outward normal vector to a supporting hyperplane of D at $\alpha(t)$.
- If $t_i \leq t \leq t_{i+1}$, then $\alpha(t) = p_1 g_\alpha(t + \sum_{j=1}^i w_j)$.
- For all $0 \leq s \leq w_i$, we have $p_1 g_\alpha(s + t_i + \sum_{j=1}^{i-1} w_j) = \alpha(t_i)$.
- As s varies from 0 to w_i, angle $p_2 g_\alpha(s + t_i + \sum_{j=1}^{i-1} w_j)$ varies over the range of supporting hyperplanes of D at $\alpha(t_i)$ (which may be a single angle if $\alpha(t_i)$ is not a vertex of ∂D).

Define $g_\beta: [0, 1 + w] \to \partial D \times S^1$ similarly (with α replaced everywhere by β). Note that $g_\alpha(0) = g_\beta(0)$, that $g_\alpha(1 + w) = g_\beta(1 + w)$, and that $p_2 g_\alpha$ and $p_2 g_\beta$ wrap in opposite directions around S^1. Hence for some $s \in [0, 1 + w]$ the supporting hyperplanes corresponding to $g_\alpha(s)$ and $g_\beta(s)$ will be parallel and on opposite sides of D. It follows that

$$\mathrm{SC}(D) + \epsilon \geq d_L(\alpha, \beta) \geq d(p_1 g_\alpha(s), p_1 g_\beta(s)) \geq w(D).$$

Since this is true for all $\epsilon > 0$, we have $\mathrm{SC}(D) \geq w(D)$. $\square$

The paper [22] shows that for $D \subseteq \mathbb{R}^2$ a convex polygonal domain with n vertices, the width and hence the sweeping cost of D can be computed in time $O(n)$ and space $O(n)$ using the *rotating calipers* technique.

---

[2]This is not quite precise if ∂D contains a straight line segment of nonzero length parallel to v (there are at most two such segments). In this case, pick an arbitrary point on each line segment parallel to v; each such point will be either the starting point or the ending point for both α and β.

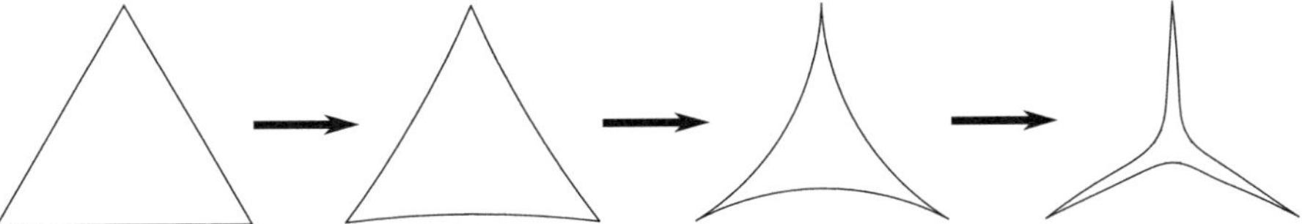

Fig. 10 We deform each edge of the triangle toward the center of the triangle, producing a three-pronged shape. As the sweeping costs of the shapes converge to a fixed constant, the areas converge to zero

9 Extremal Shapes

Which convex shape of unit area has the largest sweeping cost? The papers [11, Theorem 4.3] and [27] state that if D is a bounded planar convex domain, then area$(D) \geq w(D)^2/\sqrt{3}$, where equality is achieved if D is an equilateral triangle. The next corollary follows immediately from Theorem 8.1.

Corollary 9.1 *Let D be a convex Jordan domain. Then*

$$\text{area}(D) \geq \frac{\text{SC}(D)^2}{\sqrt{3}},$$

where equality is achieved if D is an equilateral triangle. Hence the equilateral triangle has the maximal sweeping cost over all planar convex domains of the same area.

The next example shows that there is no extremal shape for non-convex Jordan domains.

Example 9.2 A (non-convex) domain D of unit area may have arbitrarily large sweeping cost.

Proof Consider a deformation of an equilateral triangle with unit side lengths where we deform each edge toward the center of the triangle (Fig. 10). Note that as the sweeping cost converges to $\frac{1}{\sqrt{3}}$ (the distance from the center to a vertex) from above, the area of the shape tends zero. Rescaling each shape in this deformation to have area one shows that a non-convex domain of unit area may have arbitrarily large sweeping cost.

$\square$

10 Conclusion

Given a Jordan domain D in the plane, we show that the sweeping cost of D is at least as large as the shortest area-bisecting curve in D, and we give a formula for the sweeping cost in terms of the geodesic Fréchet distance between two curves on the boundary of D with non-equal winding numbers. We show that the sweeping cost of any convex domain is equal to its width. Therefore, the sweeping cost of a polygonal convex domain with n vertices can be computed in time and space $O(n)$, and a convex domain of unit area with maximal sweeping cost is the equilateral triangle.

We end by mentioning two related settings of interest. First, let D be a compact region in the plane, perhaps not simply-connected. Suppose the pursuer is now a union of curves. What can one say about the sweeping cost of D, measured as the sum of the curve lengths? Second, let $D \subset \mathbb{R}^n$ be the homemorphic image of the closed n-dimensional ball. What are the properties of the sweeping cost of D, when swept by an $(n-1)$-dimensional "sensor surface"? For example, what is a three-dimensional convex body of unit volume which maximizes this higher-dimensional sweeping cost?

Appendix: Additional Lemmas and Proofs

Lemma 1 *If (X, μ) is a measure space and U_i is a sequence of measurable sets in X, then*

(1) $\mu(\liminf_i U_i) \leq \liminf_i \mu(U_i)$, *and*
(2) $\mu(\limsup_i U_i) \geq \limsup_i \mu(U_i)$ *if $\mu(X) < \infty$.*

Proof Recall $\limsup_i U_i = \bigcap_{i=1}^{\infty}(\bigcup_{j=i}^{\infty} U_i)$. Since the $\bigcup_{j=i}^{\infty} U_i$ are a decreasing sequence of sets, and since $\mu(X) < \infty$, [16, Proposition 1.2.3] implies $\mu(\limsup_i U_i) = \lim_i \mu(\bigcup_{j=i}^{\infty} U_i)$. Since $U_i \subseteq \bigcup_{j=i}^{\infty} U_j$, we have $\mu(U_i) \leq \mu(\bigcup_{j=i}^{\infty} U_j)$, and hence

$$\mu(\limsup_i U_i) = \lim_i \mu(\bigcup_{j=i}^{\infty} U_i) \geq \limsup_i \mu(U_i),$$

giving (2). The proof of (1) is similar except that the finiteness assumption is unnecessary. $\qquad\square$

Lemma 2 *Let $\mathbb{D} = \{(x, y) \in \mathbb{R}^2 \mid x^2 + y^2 \leq 1\}$ be the unit disk. Given any point $p \in \mathbb{R}^2 \setminus \mathbb{D}$ and any curve $\alpha \colon I \to \partial\mathbb{D}$, there exists a continuous function $g \colon I \times I \to \mathbb{R}^2$ such that*

- $g(0, t) = p$ *for all $t \in I$,*
- $g(1, t) = \alpha(t)$ *for all $t \in I$, and*
- $g(s, t) \notin \mathbb{D}$ *for $s < 1$.*

Proof Given $r \geq 0$, let $\alpha_r(t) \colon I \to \mathbb{R}^2$ be defined by $\alpha_r(t) = (1+r)\alpha(t)$. Fix some $\epsilon > 0$. Pick a single curve $\gamma \colon I \to \mathbb{R}^2 \setminus \mathbb{D}$ with $\gamma(0) = p$ and $\gamma(1) = \alpha_\epsilon(0)$; this is possible since $\mathbb{R}^2 \setminus \mathbb{D}$ is connected by the Jordan curve theorem. We define the function $g \colon I \times I \to \mathbb{R}^2$ as follows:

$$g(s, t) = \begin{cases} \gamma(3s) & \text{if } 0 \leq s < \frac{1}{3} \\ \alpha_\epsilon((3s - 1)t) & \text{if } \frac{1}{3} \leq s < \frac{2}{3} \\ \alpha_{(3-3s)\epsilon}(t) & \text{if } \frac{2}{3} \leq s \leq 1. \end{cases}$$

Note that g is continuous and satisfies all of the required conditions. $\qquad\square$

Acknowledgements We would like to thank Clayton Shonkwiler for pointing us to the references [11, 27].

References

1. H. Adams, G. Carlsson, Evasion paths in mobile sensor networks. Int. J. Robot. Res. **34**(1), 90–104 (2015)
2. L. Alonso, A.S. Goldstein, E.M. Reingold, "Lion and man": upper and lower bounds. ORSA J. Comput. **4**, 447–452 (1992)
3. B. Alspach, Searching and sweeping graphs: a brief survey. Le matematiche **59**, 5–37 (2006)
4. M. Berger, *Geometry I* (Springer Science & Business Media, New York, 2009)
5. F. Berger, A. Gilbers, A. Grüne, R. Klein, How many lions are needed to clear a grid? Algorithms **2**(3), 1069–1086 (2009)
6. A. Beveridge, Y. Cai, Pursuit-evasion in a two-dimensional domain. ARS Mat. Contemp. **13**, 187–206 (2017)
7. D. Bienstock, Graph searching, path-width, tree-width and related problems (a survey). DIMACS Ser. Discret. Math. Theor. Comput. Sci. **5**, 33–49 (1991)
8. R.L. Bishop, The intrinsic geometry of a Jordan domain, International Electronic Journal of Geometry, **1**, 33–39 (2018)
9. R.D. Bourgin, P.L. Renz, Shortest paths in simply connected regions in $\mathbb{R}^2$. Adv. Math. **76**(2), 260–295 (1989)
10. D. Burago, Y. Burago, S. Ivanov, *A Course in Metric Geometry*, vol. 33 (American Mathematical Society, Providence, 2001)
11. A. Cerdán, Comparing the relative volume with the relative inradius and the relative width. J. Inequal. Appl. **2006**(1), 1–8 (2006)
12. E.W. Chambers, Y. Wang, Measuring similarity between curves on 2-manifolds via homotopy area, in *Proceedings of the Twenty-Ninth Annual Symposium on Computational Geometry* (ACM, New York, 2013), pp. 425–434
13. E.W. Chambers, D. Letscher, T. Ju, L. Liu, Isotopic Fréchet distance, in *CCCG* (2011)
14. J.-C. Chin, Y. Dong, W.-K. Hon, C.Y.-T. Ma, D.K.Y. Yau, Detection of intelligent mobile target in a mobile sensor network. IEEE/ACM Trans. Netw. **18**(1), 41–52 (2010)
15. T.H. Chung, G.A. Hollinger, V. Isler, Search and pursuit-evasion in mobile robotics. Auton. Robot. **31**(4), 299–316 (2011)
16. D.L. Cohn, *Measure Theory*, vol. 165 (Springer, Berlin, 1980)
17. J. Cortes, S. Martinez, T. Karatas, F. Bullo, Coverage control for mobile sensing networks, in *Proceedings of the IEEE International Conference on Robotics and Automation*, vol. 2 (2002), pp. 1327–1332
18. V. de Silva, R. Ghrist, Coordinate-free coverage in sensor networks with controlled boundaries via homology. Int. J. Robot. Res. **25**(12), 1205–1222 (2006)
19. A. Efrat, L.J. Guibas, S. Har-Peled, J.S.B. Mitchell, T.M. Murali, New similarity measures between polylines with applications to morphing and polygon sweeping. Discret. Comput. Geom. **28**(4), 535–569 (2002)

20. L. Esposito, V. Ferone, B. Kawohl, C. Nitsch, C. Trombetti, The longest shortest fence and sharp Poincaré–Sobolev inequalities. Arch. Ration. Mech. Anal. **206**(3), 821–851 (2012)
21. B.T. Fasy, S. Karakoc, C. Wenk, On minimum area homotopies of normal curves in the plane. arXiv:1707.02251 (2017, Preprint)
22. M.E. Houle, G.T. Toussaint, Computing the width of a set. IEEE Trans. Pattern Anal. Mach. Intell. **10**(5), 761–765 (1988)
23. I. Johnstone, D. Siegmund, On Hotelling's formula for the volume of tubes and Naiman's inequality. Ann. Stat. **17**, 184–194 (1989)
24. J.R. Kline, Separation theorems and their relation to recent developments in analysis situs. Bull. Am. Math. Soc. **34**(2), 155–192 (1928)
25. B. Liu, P. Brass, O. Dousse, P. Nain, D. Towsley, Mobility improves coverage of sensor networks, in *Proceedings of the 6th ACM International Symposium on Mobile Ad hoc Networking and Computing* (ACM, New York, 2005), pp. 300–308
26. Z. Nie, On the minimum area of null homotopies of curves traced twice. arXiv:1412.0101 (2014, Preprint)
27. J. Pál, Ein minimumproblem für ovale. Math. Ann. **83**(3), 311–319 (1921)
28. T.D. Parsons, Pursuit-evasion in a graph, in *Theory and Applications of Graphs* (Springer, Berlin, 1978), pp. 426–441
29. G. Pólya, *Mathematics and Plausible Reasoning: Volume 1: Induction and Analogy in Mathematics* (Oxford University Press, Oxford, 1965)
30. E. Sriraghavendra, K. Karthik, C. Bhattacharyya, Fréchet distance based approach for searching online handwritten documents, in *Ninth International Conference on Document Analysis and Recognition*, vol. 1 (2007), pp. 461–465
31. G.T. Toussaint, Solving geometric problems with the rotating calipers, in *Proceedings of IEEE Melecon*, vol. 83 (1983), p. A10
32. C. Wenk, A.F. Cook IV, Geodesic Fréchet distance inside a simple polygon. ACM Trans. Algorithms **7**(1), 9 (2010)
33. L. Zoretti, Sur les fonctions analytiques uniformes. J. Math. pures appl **1**, 9–11 (1905)

Scaffoldings and Spines: Organizing High-Dimensional Data Using Cover Trees, Local Principal Component Analysis, and Persistent Homology

Paul Bendich, Ellen Gasparovic, John Harer, and Christopher J. Tralie

Abstract We propose a flexible and multi-scale method for organizing, visualizing, and understanding point cloud datasets sampled from or near stratified spaces. The first part of the algorithm produces a cover tree for a dataset using an adaptive threshold that is based on multi-scale local principal component analysis. The resulting cover tree nodes reflect the local geometry of the space and are organized via a *scaffolding* graph. In the second part of the algorithm, the goals are to uncover the strata that make up the underlying stratified space using a local dimension estimation procedure and topological data analysis, as well as to ultimately visualize the results in a simplified *spine* graph. We demonstrate our technique on several synthetic examples and then use it to visualize song structure in musical audio data.

1 Introduction

We consider *point cloud data*, modeled as a set of points $\mathbb{X} = \{x_1, \ldots, x_m\}$ in a Euclidean space $\mathbb{R}^D$. Such clouds are hard to analyze directly when m and/or D is large. Subsampling techniques are often used in the former situation, and dimension

P. Bendich
Department of Mathematics, Duke University, Durham, NC, USA
Geometric Data Analytics, Inc., Durham, NC, USA
e-mail: bendich@math.duke.edu

E. Gasparovic (✉)
Department of Mathematics, Union College, Schenectady, NY, USA
e-mail: gasparoe@union.edu

J. Harer
Department of Mathematics, Duke University, Durham, NC, USA
Department Electrical and Computer Engineering, Duke University, Durham, NC, USA
Geometric Data Analytics, Inc., Durham, NC, USA
e-mail: harer@math.duke.edu

C. J. Tralie
Department of Electrical and Computer Engineering, Duke University, Durham, NC, USA
e-mail: cjt16@duke.edu

© The Author(s) and the Association for Women in Mathematics 2018
E. W. Chambers et al. (eds.), *Research in Computational Topology*, Association
for Women in Mathematics Series 13, https://doi.org/10.1007/978-3-319-89593-2_6

reduction in the latter. In both cases, much care has to be taken to ensure that the reduction in the number of points and number of dimensions does not destroy essential features of the data. While theorems exist in both contexts, they tend to make assumptions about parameters (like intrinsic dimension) that may be unknown or that may vary widely across $\mathbb{X}$. The latter problem often occurs when $\mathbb{X}$ is not sampled from a manifold, but rather from a *stratified space*.

A stratified space is a topological space that can be decomposed into manifold pieces (called *strata*), of possibly different dimension, all of which fit together in some uniform fashion. A key distinction is between *maximal* and *non-maximal*, or *singular* strata: a maximal stratum does not belong to the closure of any other stratum, and a non-maximal stratum occurs where two or more maximal strata meet. See Sect. 2.3 and see Fig. 2 for an example.

This paper proposes a novel, flexible, and fast new technique for high-dimensional data organization, visualization, and analysis. The most interesting datasets to apply this method to are those sampled from stratified spaces. The method involves a data structure called a *cover tree* [5] and employs techniques derived from *multi-scale local principal component analysis* (MLPCA, [16]) and *topological data analysis* (TDA, [12]).

Our method uncovers and summarizes the strata that make up the underlying stratified space, exhibits how the different pieces fit together, and reflects the local geometric and topological properties of the space. We derive from $\mathbb{X}$ a multi-scale set of graphs, called the *scaffoldings* and the *spine* of $\mathbb{X}$, that enable one to visualize the stratified structure of the data. At each fixed scale, a point in $\mathbb{X}$ belongs to a unique node in these graphs. A subset of points belongs to a common node only if the local geometry at each of those points is similar enough; that is, only if they belong to a common stratum. We also determine whether a node corresponds to a maximal or non-maximal stratum. In the former case, this suggests that a single-dimension reduction technique might be employed on the points in that node. The edges of these graphs give information about the possible transition regions between different zones of local similarity.

1.1 Outline of Method and Paper

After a brief discussion of relevant graph theory (Sect. 2.1), we explain how to represent a point cloud $\mathbb{X}$ by a cover tree $\mathscr{T}$, which is a rooted tree structure with an associated collection of subsets of $\mathbb{X}$ called "levels." The points in each level, called "centers" or "nodes," are evenly distributed and representative of the point cloud at that scale. See Sect. 2.2 for more details.

For the next step, described in detail in Sect. 3, we use MLPCA (Sect. 2.4) to select a subset of nodes $\mathscr{V}$ from $\mathscr{T}$. Each node in $\mathscr{V}$ represents a collection of points for which the local geometric structural information (as measured by the eigenvalues of covariance matrices) remains relatively constant. The nodes in $\mathscr{V}$ form the vertices of a graph Σ, whose edge set $\mathscr{E}$ is computed via the geometry of the ambient space.

Then we label each node $v \in \mathscr{V}$ with an estimated local dimension $\mathscr{F}(v)$, using insights from MLPCA and the theory of stratified spaces; this is described in Sect. 4. Formally, the *scaffolding* of $\mathbb{X}$ is the triple $\Sigma = (\mathscr{V}, \mathscr{E}, \mathscr{F})$, with $\mathscr{F} : \mathscr{V} \to \{0, 1, 2, \ldots\}$. See Figs. 6d and 8a for examples. From Σ, we study the local topological structure of the dataset using persistent homology (Sect. 2.5).

One may then (Sect. 5) use graph-collapsing techniques, again informed by insights from the theory of stratified spaces, to produce a much smaller graph $S = (V, E, F)$, called the *spine* of $\mathbb{X}$. Here a subset of points belongs to a fixed node in V only if those points belong to the same stratum. See Fig. 8b for a simple example of a spine. The end product of the algorithm is an efficient summary and visualization of the strata that have been determined to make up the underlying stratified space. Thus, the flow of the algorithm may be summarized as follows:

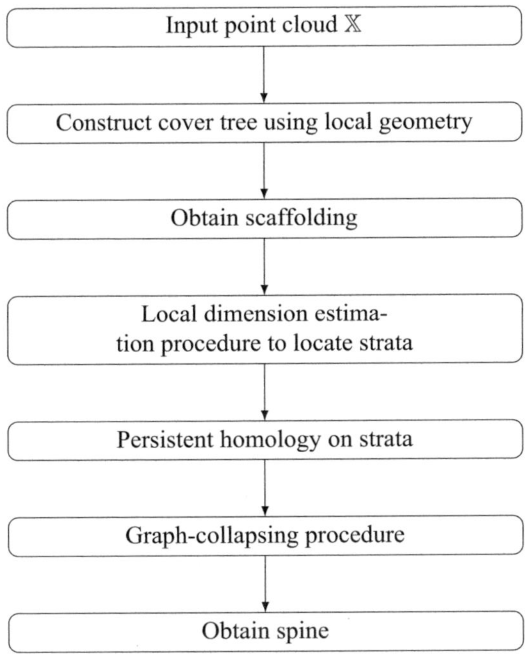

To illustrate this framework, we first run several experiments (Sect. 6) on low-dimensional synthetic datasets, including two that have already been studied in the literature, whose goal is to provide a means of organizing and visualizing the underlying stratified structure of the dataset. We also give high-dimensional examples of visualization experiments with real data, constructing scaffoldings and spines for point clouds created by time series embeddings of musical audio data.

1.2   Related Work

MLPCA has been used before [2, 4] to analyze data, but this is the first work to use it in combination with the cover tree.

The cover tree was first introduced in [5] as a way of performing approximate nearest neighbors in $O(\log(N))$ time for a point cloud with N points in arbitrary metric spaces of a fixed dimension (though constants depend exponentially on the dimension). More recent work has shown that cover trees can be used for analysis of high-dimensional point clouds with a low-dimensional intrinsic structure [10]. Similarly to [10], we use the cover tree to deconstruct the point cloud into geometrically simpler parts, but we encode more explicitly the stratified space structure in our representation, and we also autotune which nodes at which levels represent each part, as explained in Sect. 3.

Many papers (e.g., [14]) have analyzed stratified spaces as mixtures of general manifolds, and one [23] as a union of flats. Using techniques derived mostly from Gaussian Mixture Models in the former, and Grassmanian methods in the latter, they prove theoretical guarantees about when a point belongs to a maximal stratum, but they do not say much about the singularities. Other work [3] uses topological data analysis to make theoretically-backed inferences about the singularities themselves; however, the algorithms are too slow to be practical. Finally, a general survey of subspace clustering algorithms may be found in, for example, [20].

2   Background

2.1   Graph Theory

We give the basic vocabulary about graphs and rooted trees that we will need, the former for defining the spine graph and the latter for the cover tree. For the expert, we note that our graphs are undirected and simple.

A *graph* $G = (V, E)$ consists of a set of *vertices* (or nodes) V and a collection of *edges* E, where each edge $e \in E$ is an unordered pair $e = \{x, y\}$ of vertices. Given such an edge, we say that x and y are *adjacent*, that e *connects* x and y, and that x and y are both *incident* to e. The *link* $L(x)$ of a vertex x consists of all vertices adjacent to x, and the link of $e = \{x, y\}$ is $L(e) = L(x) \cap L(y)$. Suppose f is a real-valued function defined on V. The *upper link* of a vertex $x \in V$ is $L_+(x) = \{y \in L(x) \mid f(y) > f(x)\}$, and the upper link of the edge e is $L_+(e) = L_+(x) \cap L_+(y)$.

Whenever W is a subset of V, we can form a new graph $G' = (V', E')$ by *deleting* W. Here $V' = V - W$, and $\{x, y\} \in E'$ whenever x and y were both adjacent to some vertex in W; all edges from E that connected two non-W vertices are retained.

Alternatively, we can form a different graph $G'' = (V'', E'')$ from G by *collapsing* W. In this case, V'' is obtained from V by removing W and then adding in a dummy vertex w. The edges E'' are obtained from E by deleting all edges that connect pairs of vertices from W, and also drawing a new edge from w to each $x \in V - W$ such that x was connected to some vertex from W in G.

A *path* of length m between x and v in G is a collection of distinct vertices $x = v_0, \ldots, v_m = v$ such that each v_i and v_{i+1} are adjacent. A *cycle* is a path from a vertex back to itself. If there exists a path in G between every pair of vertices, then G is *connected*. Otherwise, G can be decomposed into two or more *connected components*.

A *tree* T is a connected graph with no cycles. If an arbitrary vertex $v \in T$ is distinguished as a *root*, then T becomes a *rooted tree*. Each non-root node x has a unique path to the root. The mth *level* of T consists of all vertices for which the length of this path is m, and we call T a *leveled tree*. We define a partial ordering on the vertices by declaring $y \leq x$ if and only if the path from y to v passes through x, in which case we say that y is a *descendant* of x and that x is an *ancestor* of y. If y and x are one step away from each other, then we use *child* and *parent* instead.

2.2 Cover Tree

In this section, we give the definition of a *cover tree* associated to a point cloud $\mathbb{X}$ as in [5]. We use $B_{R_i}(x)$ to denote the ball of radius R_i centered at the point $x \in \mathbb{X}$, and in what follows, we take $R_i = 1/2^i$ for $i \in \mathbb{Z}$. The given dataset $\mathbb{X}$ is normalized to the unit ball prior to computing the cover tree.

Definition 1 A **cover tree** $\mathcal{T}$ on a point cloud $\mathbb{X}$ with a metric $d_{\mathbb{X}}$ is a leveled tree on $\mathbb{X}$, where each node at each level is associated to one of the points in $\mathbb{X}$ (called "centers"), and where the collection of C_i (set of points in $\mathbb{X}$ that are level i centers for $i \in \mathbb{Z}$) satisfy the following three conditions:

- (Nesting) $C_i \subset C_{i+1}$.
- (Covering): For each $C_{i+1}^j \in C_{i+1}$, there exists a node $y \in C_i$ so that $d_X(C_{i+1}^j, y) \leq R_i$.
- (Packing): For all $C_i^j \neq C_i^k \in C_i$, $B_{R_i}(C_i^j) \cap B_{R_i}(C_i^k) = \emptyset$.

Here, C_i^j indicates the jth center at level i.

Another way of phrasing the covering condition is to say that balls of radius R_i centered at the points in C_{i+1} will cover the point cloud. Exactly one such $y \in C_i$ is designated as the parent of C_{i+1}^j in the tree structure. The packing condition promotes levels which contain equally spaced centers. Note that in this definition, it is possible for a node at one level to be its own parent one level up. There is a minimum i for which the size of C_i is 1, and the node at that value of i is designated as the root node. Similarly, there is a maximum i for which $C_i = \mathbb{X}$. Figure 1 shows an example cover tree on a point cloud in $\mathbb{R}^2$.

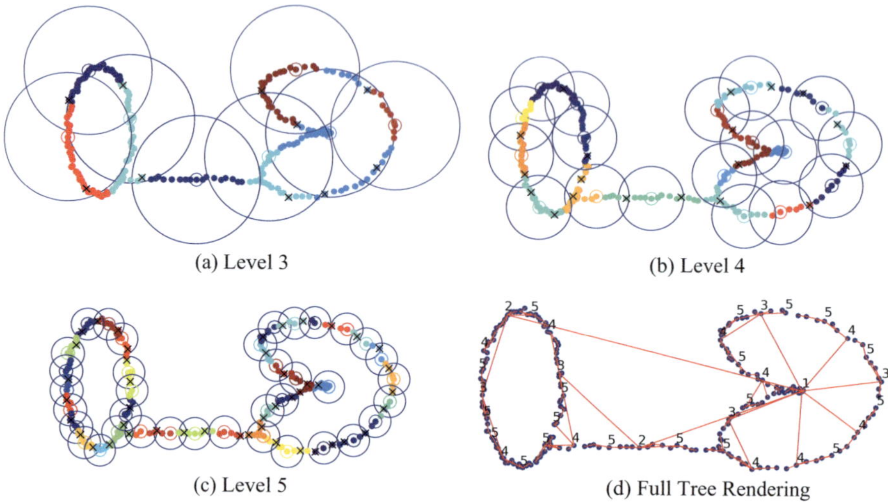

(a) Level 3   (b) Level 4

(c) Level 5   (d) Full Tree Rendering

Fig. 1 An example of a cover tree on 250 points in $\mathbb{R}^2$. In (**a**)–(**c**), the centers at level i are circled, and disks with radius $1/2^i$ are drawn in blue around each center. Points within the same subtree rooted at a center at that level are drawn with the same color. Centers at level $i + 1$ are marked with an X to illustrate the covering condition. The tree is drawn in (**d**), with points in the first five levels numbered with the level in which they first occur

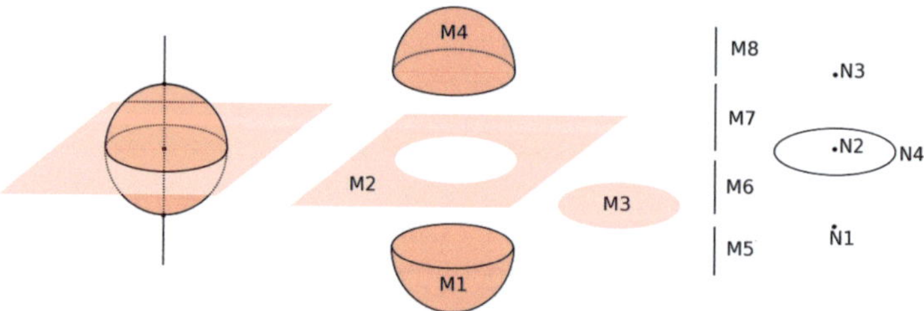

Fig. 2 Far left: A simple stratified space Y. Rest of figure: a stratification of Y into 12 strata; M_1 through M_8 are maximal, and N_1 through N_4 are non-maximal

2.3 Stratified Spaces

Recall that a space Y is called a *manifold* of dimension k if every point in Y has a neighborhood that is homeomorphic to $\mathbb{R}^k$. The space Y on the left of Fig. 2, consisting of a sphere intersecting with a horizontal plane at its equator and with a line piercing it at the north and south poles, is not a manifold: For example, every small enough neighborhood of the north pole looks like a plane pierced by a line. A *stratification* of Y is a decomposition of Y into connected manifolds $\{S_i\}$ called *strata*. The strata must satisfy certain requirements for fitting together within the whole of Y; see, e.g., [15]. The rest of the figure shows a stratification of Y.

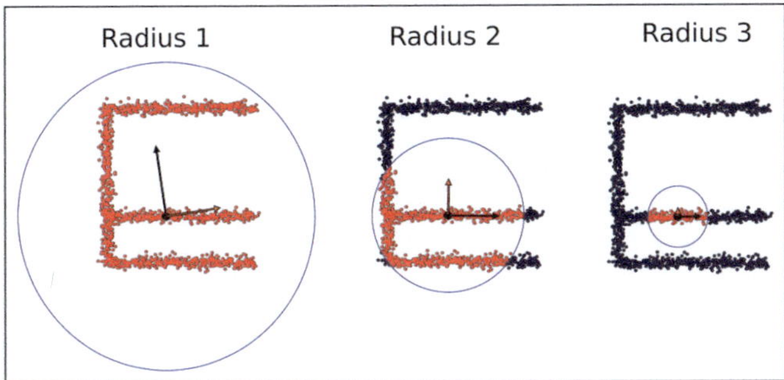

Fig. 3 Illustration of the technique of multi-scale local principal component analysis (MLPCA)

A stratum S_i is called *maximal* if it is disjoint from the closure of any other stratum. The other strata are called *non-maximal*. If z is a point on a non-maximal stratum, then every small neighborhood of it within Y looks like the intersection of two or more strata. In our example, the maximal strata are labeled M_1 through M_8; note that there is no reason for the maximal strata to all have the same dimension. Each of the non-maximal strata, N_1 through N_4, belongs to the closure of two or more of the maximal strata.

2.4 Local Principal Component Analysis

We assume the reader is already familiar with principal component analysis (PCA) [22]. Here we regard PCA simply as a machine that takes in a point cloud, computes a covariance matrix, and returns the eigenvectors along with the corresponding eigenvalues given in non-increasing order. This "eigeninformation" may then be used for dimensionality reduction.

The technique of *multi-scale local principal component analysis* (MLPCA) [16], takes as input a point cloud $\mathbb{X} \subset \mathbb{R}^D$, a particular point $p \in \mathbb{R}^D$, and a radius R, and returns as output the results of PCA run only on the points in $\mathbb{X}$ that lie within the Euclidean R-ball around p. This is typically done at multiple radius scales and at many center points; see Fig. 3.

We close with an observation that will be crucial later. Suppose the point cloud $\mathbb{X}$ has been sampled from a stratified space Y. After computing MLPCA at a few radii around a specific point $p \in \mathbb{X}$, one might consistently see k dominant eigenvectors, and thus try to conclude that the "local dimension" around p is k; in other words, that Y looks locally like a k-flat near p. This is certainly possible but need not be the case. Instead, p might be near a non-maximal stratum of Y, and the local picture might be more like the intersection of an ℓ-flat and a $k - \ell$ flat, for some value of ℓ.

2.5 *Persistent Homology*

Algebraic topology associates to a topological space X a series of abelian groups $H_i(X)$ called the *homology groups* of X which measure its higher-order connectivity. The ranks of these groups are called the *Betti numbers $\beta_i(X)$*: β_0 counts the number of connected components of X, β_1 (resp. β_2) counts the number of independent non-bounding loops (resp., closed surfaces) in the space, and so forth. See [19], for example, for precise definitions.

Persistent homology transforms these algebraic notions into a measurement tool relevant to high-dimensional and noisy data. The key idea is that we are rarely interested in the Betti numbers of a particular fixed space; rather, we are concerned with the homological features that persist across a wide interval in a one-parameter filtration of a space, and we represent these features in a compact visual way as a *persistence diagram* in the plane. See, e.g., [12] or [9].

The simplest homological invariant of a space X is $\beta_0(X)$, its number of components. Let $\mathbb{X}$ be the point cloud seen on the left of Fig. 4. The most literal interpretation of $\beta_0(\mathbb{X})$ here would be N, where $N = |\mathbb{X}|$. On the other hand, most off-the-shelf clustering algorithms would report that $\mathbb{X}$ has three components, since that seems to be the most natural way to group the points. Some algorithms might report back two clusters (grouping the top two into one), depending on how parameters are chosen. Persistent homology in dimension zero gives a useful description of all of these answers along with a report of the scales at which they are valid.

More precisely, let us define, over a set of *scales* $\alpha \geq 0$, the space $\mathbb{X}_\alpha$ to be the union of closed balls of radius α centered at each point in the cloud. As α increases, we imagine the point cloud gradually thickening into the ambient space, and more and more components begin to merge into one another. An α-value at which a merger occurs will be called a *death*, and we record the multi-set of these deaths as dots in the zero-dimensional *persistence diagram* of $\mathbb{X}$, as shown on the

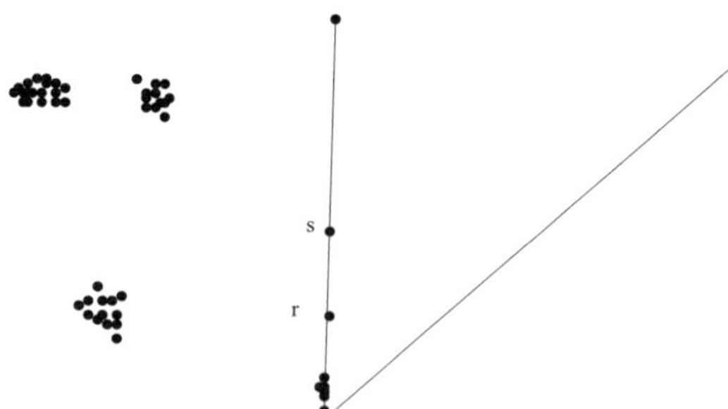

Fig. 4 A point cloud $\mathbb{X}$ (left) and the zero-dimensional persistence diagram (right). The x- and y-axes represent birth and death time of components. The values r and s are half the distance between the two top clusters and between the right and bottom clusters, respectively

Fig. 5 A point cloud sampled from a closed curve in the plane (left) and the persistence diagram in dimension one (right). Again, the axes represent birth and death time of loops

right side of Fig. 4. One way to translate between this diagram and a traditional number-of-clusters answer is to cut the diagram by a horizontal line, and then report the number of dots above that line. This shows that the answers "two clusters" and "three clusters" are both quite reasonable, since each could be obtained via a wide range of threshold parameters. On the other hand, the answer "nine clusters" is in no way stable, since there is only a very narrow range of threshold parameters that would report that number.

In general, the k-dimensional persistence diagram $D_k(\mathbb{X})$ of a point cloud will summarize the evolution of the k-dimensional homology group $H_k(\mathbb{X}_\alpha)$, while $\mathbb{X}$ thickens as described above. For instance, consider the left side of Fig. 5, which shows a point cloud $\mathbb{X}$ sampled, with some noise, from a closed curve in the plane. As $\mathbb{X}$ starts to thicken, all of the components very quickly collapse into one. However, there is also a new phenomenon that occurs: namely, the formation of a closed loop. We call this the *birth* of a one-cycle. When α increases enough to build a bridge across the neck of the shape, a new one-cycle is born. Each of these one-cycles is said to *die* at the α-value for which it is possible to contract the loop to a point within $\mathbb{X}_\alpha$. The (birth, death) pairs corresponding to the lifetime of each cycle are stored in $D_1(\mathbb{X})$, shown on the right side of the same figure. The *persistence* of a dot $u = (x, y)$ is $y - x$.

Suppose that $\mathbb{X}$ is a good sampling from a compact space Y. There are then a series of "homology inference" theorems [9, 11] which prove that one can read off the "ground truth" homology of Y by examining the points in $D_k(\mathbb{X})$ which sit near the y-axis and are of high-enough persistence.

3 The Adaptive Cover Tree and the Scaffolding

The goal of the first part of our algorithm is to break up a point cloud $\mathbb{X} \subseteq \mathbb{R}^D$ (sampled from or near a connected stratified space) into clusters such that the points in a given cluster are sampled from the same stratum. One could let $d_{\mathbb{X}}$ be the Euclidean metric and build the full cover tree $\mathscr{T}$ as above. Instead, at each node in the cover tree, we use a criterion to decide whether we will continue to construct the subtree rooted at that node, or whether all of the points contained in that subtree are sufficiently "uniform." In the latter case, the entire subtree is compressed to a single node. At the end of this adaptive process, we output the leaves as $\mathcal{V}$, the nodes of our scaffolding graph Σ.

3.1 Eigenmetric Threshold

Our adaptive tree building construction is based on MLPCA, as we now explain. First, for some increasing set of radii $r_1, \ldots r_n$ and for each point $p \in \mathbb{X}$, we let $\boldsymbol{\lambda}^i(p) = (\lambda_1^i, \lambda_2^i, \ldots \lambda_D^i), i = 1, \ldots, n$ be the eigenvalues, in non-increasing order, that result from performing MLPCA on $\mathbb{X}$ with radius r_i and center point p. Then for any pair of points $p, q \in \mathbb{X}$, we define

$$E(p, q) = ||(||\boldsymbol{\lambda}^1(p) - \boldsymbol{\lambda}^1(q)||, \ldots, ||\boldsymbol{\lambda}^n(p) - \boldsymbol{\lambda}^n(q)||)||, \tag{1}$$

where $|| \cdot ||$ is the usual Euclidean norm. We shall refer to E as the *eigenmetric* on $\mathbb{X}$. Points with similar eigenvalues across multiple scales will have small eigenmetric distance. Note that E is a pseudometric as $E(p, q)$ may be 0 for $p \neq q$ (e.g., consider two copies of the same point cloud laid out side by side).

The eigenmetric is used to construct $\mathcal{V}$ as follows. Fix a level i of $\mathcal{T}$. At each node C_i^j on this level, we take evenly spaced radii up to the radius of the node and we perform an eigenmetric threshold test on the points from $\mathbb{X}$ associated with C_i^j that would end up in its subtree upon further construction. To do this, we look at the points associated with each child of C_i^j, and we only further subdivide a node when the eigenmetric between a pair of children exceeds an *eigenthreshold* τ. Otherwise, we stop constructing the subtree below C_i^j and instead add C_i^j to $\mathcal{V}$. At the end of this process, every point in $\mathbb{X}$ belongs to a unique node in $\mathcal{V}$. We refer to the set of points belonging to a given node v as the *cluster* of v.

One can specify the eigenthreshold τ as follows. If n is the number of radii, and if one desires that, for every $i = 1, \ldots, n$, $||\boldsymbol{\lambda}^i(p) - \boldsymbol{\lambda}^i(q)|| < \epsilon$ for some small value of ϵ, then one should set $\tau = \sqrt{n}\epsilon$.

Parts (a)–(c) of Fig. 6 show this process. To reap the multi-scale benefits, we construct the cover tree nodes for a range of eigenthreshold values and obtain a multi-scale set of scaffolding graphs. In every example in this paper, we always use the default value of $n = 30$ for the number of radii.

3.2 Building the Scaffolding

Now, we describe how to organize the results of the cover tree based on geometric proximity. This is achieved through the construction of the scaffolding graph Σ on the nodes $\mathcal{V}$. A pair of nodes $v, w \in \mathcal{V}$ forms an edge in $\mathcal{E}$ if and only

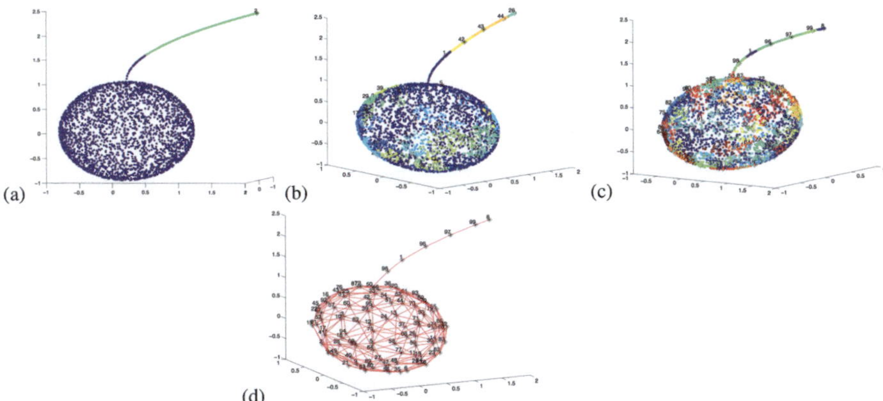

Fig. 6 Results of the adaptive cover tree construction corresponding to eigenthresholds (**a**) $\tau = 1$, (**b**) $\tau = 0.1$, and (**c**) $\tau = 0.001$ for a point cloud sampled from a sphere with an adjoining curve. Points belonging to a single $\mathcal{V}$-node are given the same color. Also shown is (**d**): A scaffolding for the last cover tree with automatically determined distance threshold $\delta \approx 0.1$

if the Euclidean distance between some pair of points in the associated clusters is below some threshold. Although the value δ of this threshold can be chosen manually, we use an automatic method for determining the threshold based on zero-dimensional persistent homology. Namely, we compute $D_0(\mathbb{X})$ to quantify how much thickening is needed before the points in $\mathbb{X}$ become a single component. This smallest thickening value, d, is the persistence of the second longest living homology class in dimension zero. Then we set the distance threshold $\delta = d$ (or better, to $d + \epsilon$ for some small value of ϵ; see Theorem 1). See part (d) of Fig. 6 and part (a) of Fig. 8 for examples of scaffolding graphs.

4 Local Dimension Estimation and Strata Determination

In the second part of our algorithm, we seek to uncover the underlying strata and determine their dimension and local topological characteristics.

4.1 Local Dimension Estimation

Once the scaffolding graph Σ is constructed, we compute a nonnegative-integer-valued function $\mathcal{F}$ on $\mathcal{V}$. If the cluster of a node v comes from well within a maximal stratum, then $\mathcal{F}(v)$ will estimate the local dimension of that stratum. We describe our choice of $\mathcal{F}$ here, but point out that any number of other intrinsic dimension estimation techniques (e.g., the box-counting method of [14]) can be substituted into our essential framework.

For each node v, we begin by performing PCA on a point set made up of the union of the cluster of v and all clusters corresponding to neighbors of v in Σ. We compute the square roots of the eigenvalues from PCA, normalizing so that their values range between 0 and 1. We then compute the differences between successive eigenvalues, including the difference between the smallest eigenvalue and 0, and use the location of the largest "eigengap" as our initial estimate of local dimension near that cluster.

Refining Dimension Estimates Next, we refine these initial estimates by building in knowledge of maximal vs. non-maximal strata in generic stratified spaces. Specifically, we consider the nodes $w \in \mathcal{V}$ which have at least one neighbor $x \in L(w)$ such that $\mathcal{F}(x) < \mathcal{F}(w)$. If w belongs to the set $\mathcal{W}$ of such nodes, then it is possible that the cluster of w was sampled from a non-maximal stratum. This is due to the fact that, when two or more maximal strata come together, PCA will see the dimension near the intersection as the sum of the dimensions of the individual maximal strata, up to the dimension of the ambient space.

Now if $w \in \mathcal{W}$ truly represents a non-maximal stratum where two or more maximal strata come together to create a singularity, then (1) its link $L(w)$ in the scaffolding should be disconnected (with the connected components corresponding to the nearby maximal strata) and (2) the link should include nodes whose dimensions ($\mathcal{F}$-values) sum to at least its estimated dimension, that is,

$$\mathcal{F}(w) \leq \sum_{x \in L(w)} \mathcal{F}(x). \tag{2}$$

If either of these conditions fails to hold, we conclude that the $\mathcal{F}$-values in $L(w)$ were miscalculated and relabel them all with $\mathcal{F}(w)$.

Figure 7 shows an example, with points sampled from a plane intersecting a line in $\mathbb{R}^3$. The nodes $\mathcal{V}$ of the scaffolding Σ are shown as black asterisks, and the local dimension estimation is illustrated by color in the figure. Red points belong to nodes labeled three-dimensional, and since such nodes have neighbors in the scaffolding with smaller dimensions, they are candidates for non-maximal strata. Green points scattered on the plane belong to nodes labeled one-dimensional, so the blue two-dimensional nodes that are connected to such one-dimensional nodes are also candidates for being non-maximal strata. After refining the estimates using the theory above, the one-dimensional nodes scattered on the plane are determined to be errors in the refining process, and are therefore relabeled with improved dimension estimates. Parts (a) and (b) show the estimates before and after refinement, respectively.

The local dimension estimation process is also a multi-scale process. It is possible that performing PCA on a node and its neighbors in the scaffolding at a small distance threshold does not give an accurate representation of the local geometric structure near the node. Perhaps there are too few points to capture the local picture, or perhaps the shape of the clusters is such that the dimension estimation from PCA would result in an error. To improve one's understanding of the intrinsic local dimension, it is best to employ a generalized version of MLPCA in the local

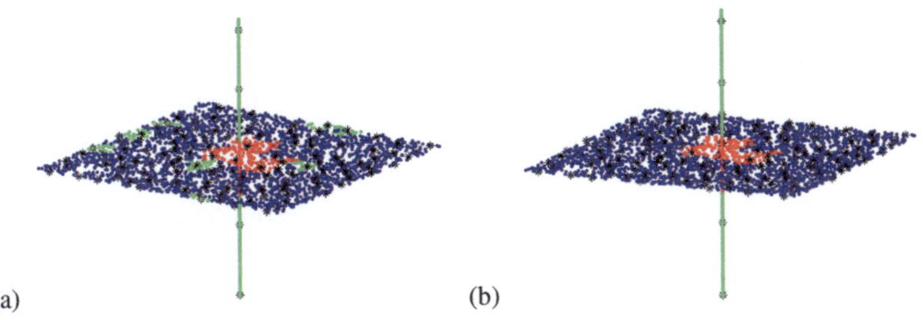

Fig. 7 Results of local dimension estimation prior to refining the estimates (**a**) and after the refinement process (**b**)

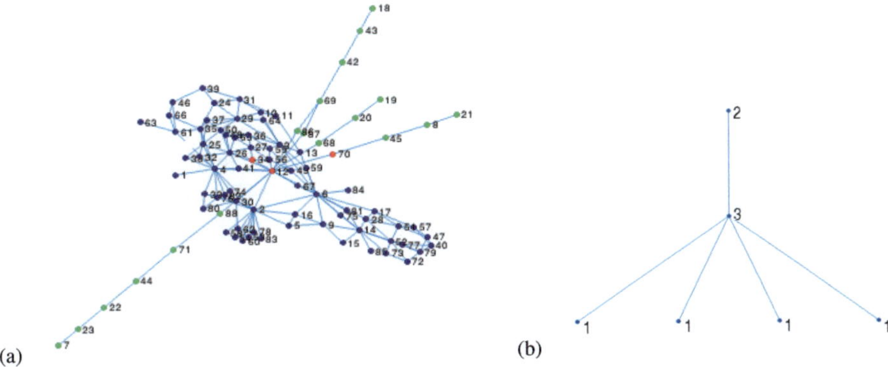

Fig. 8 For a point cloud sampled from two lines intersecting a plane, the scaffolding (**a**) and spine (**b**) are pictured. In (**a**), green, blue, and red points indicate one-, two-, and three-dimensional areas, respectively

dimension estimation process. That is, one may preserve the value of δ for the scaffolding but gradually relax the threshold for including more clusters in the PCA computations for each node. For example, for the point cloud giving rise to the scaffolding in Fig. 8a, although the distance threshold choice of $\delta = 0.05$ yields a connected scaffolding graph, the dimension estimation process of performing PCA on a cluster and those clusters connected to it in the scaffolding yields a too-local understanding of the dataset for this choice of δ. The results of the local dimension estimation that are shown in Fig. 8 were the result of computing, for each cluster, PCA on the clusters within a distance threshold range of approximately 0.12–0.23.

4.2 Topology of the Strata

To understand the topology of the strata, we perform zero-, one-, and two-dimensional persistent homology on the node centers (or, if the point cloud is not very large, on all of the points) corresponding to each dimension-based connected component in the scaffolding graph. This gives local topological information rather than the global picture one gets from ordinary persistent homology performed on the entire dataset. We decorate the nodes with vectors of persistent Betti numbers $(\beta_0, \beta_1, \beta_2)$; see Figs. 9 or 10a. A non-decorated node can be assumed to have persistent homology described by the vector $(\beta_0, \beta_1, \beta_2) = (1, 0, 0)$.

As we will see in the examples in Sect. 6, our renderings preserve essential topological features—namely, the actual loops present in the data—that a linear technique, such as PCA, will often completely destroy.

5 The Spine

Local dimension estimation works best when there are many cover tree nodes, which is a result of choosing a small eigenthreshold. Consequently, the scaffolding Σ has many vertices. As a final step in our algorithm, we now describe a procedure that radically simplifies Σ to produce a new graph S on a much smaller vertex set. This graph, called the *spine* of $\mathbb{X}$, provides a streamlined and efficient summary of the strata that make up the stratified space from which $\mathbb{X}$ was sampled.

Using the procedure above, $\mathcal{V}$ is partitioned into maximal nodes $\mathcal{M}$ and non-maximal ones $\mathcal{N}$. We then further partition $\mathcal{N}$ by $\mathcal{F}$-value and locate the connected components therein. These connected components are then collapsed into single nodes (as described in Sect. 2.1). Note this retains all edges that previously connected these components to non-maximal nodes of other dimensions, as well as to maximal nodes.

Now we process the maximal nodes $\mathcal{M}$, starting by partitioning them by $\mathcal{F}$-value and then taking connected components within this partition. Let $\mathcal{C}$ be one such component. We divide $\mathcal{C}$ further into *boundary* nodes and *interior* nodes, where a node is in the boundary if it is connected to at least one node of a different dimension. The interior nodes are deleted (as described in Sect. 2.1). Then, for any edge $e = \{x, y\}$ that connects two boundary nodes, we check whether $L_+(x) = L_+(y)$; in other words, we check whether x and y are connected to the same set of non-maximal nodes. If so, then the edge e is collapsed. Running through this process for each component of $\mathcal{M}$ produces the vertices V and edges E of the spine S. We note that vertices with different $\mathcal{F}$-value never get collapsed together. Therefore, it makes sense to label each node in V with the $\mathcal{F}$-value of any node that was collapsed into it. This labeling produces F, and we have our spine $S = (V, E, F)$. For example, the scaffolding Σ for a point cloud sampled from two lines intersecting a plane in $\mathbb{R}^3$ is pictured in Fig. 8a. If we run this process on Σ, we obtain the spine S in part (b). Note that S is in fact the *Hasse diagram* for the partial ordering on the strata of a plane pierced by two lines. We recall that the Hasse diagram for a partially ordered set $(X, <)$ [6] is a directed graph $G = (X, E)$, where there is an arc from x to y if and only if $x < y$. If $\{S_i\}$ is a stratification of a space Y, then there is a partial ordering on the set $\mathcal{S}$ of strata, where if $S_i < S_j$, the local neighborhood of a point in S_i looks like the intersection of S_j with one or more other strata.

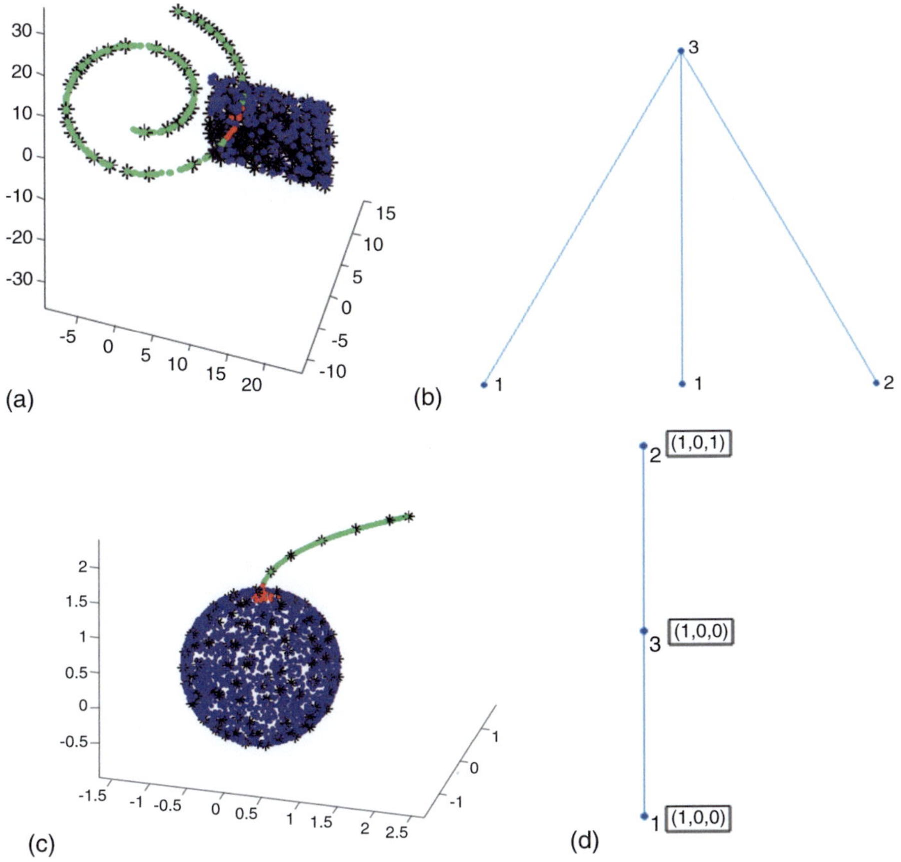

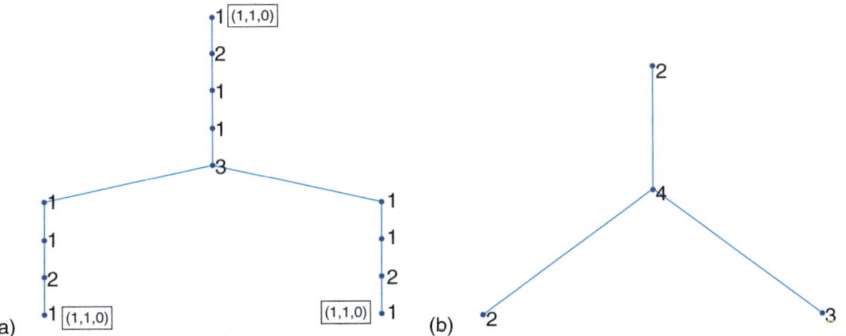

Fig. 9 Local dimension estimation and spine for the spiral-plane example ((**a**) and (**b**)) and the sphere-curve example ((**c**) and (**d**))

Fig. 10 (**a**) The spine for a point cloud sampled from three lollipop-like structures in $\mathbb{R}^6$. (**b**) The spine for a point cloud sampled from a 2-plane and 3-plane intersecting along a line in $\mathbb{R}^4$

We conclude our discussion of our algorithm with a result regarding the stability of the parameter choices.

Theorem 1 *Let Σ and S denote the scaffolding and spine graphs, respectively, constructed for the point cloud $\mathbb{X}$ for eigenthreshold τ and distance threshold δ. There exist open intervals $(\tau - \alpha, \tau + \alpha)$ and $(\delta - \beta, \delta + \beta)$ for values $\alpha, \beta > 0$ such that the scaffolding and spine graphs obtained using parameter values in these intervals are equal to Σ and S.*

Proof We show that the scaffolding graph does not change, from which it follows that the spine also will not change. First, the value of τ determines when a node is further subdivided by requiring that, for every pair of children p, q, $E(p, q) < \tau$. Clearly, there exists an open interval around τ such that, for values of the eigenthreshold in this open set, the subdivision decisions will result in the same nodes as those in Σ. Second, we assume that $\delta = d + \epsilon$ where d is obtained using persistent H_0 as in Sect. 3.2, and $2\epsilon > 0$ is the value such that a threshold of $d + 2\epsilon$ adds at least one additional edge to Σ. Then choose $0 < \beta < \epsilon$ so that $(\delta - \beta, \delta + \beta)$ is the desired open interval such that distance thresholds chosen from this interval will not add or remove edges in Σ.

We aim to prove stronger stability results in future work; see Sect. 7. Note that Theorem 1 ensures that if S is the Hasse diagram of the underlying stratified space, then for an open set of the parameters, the resulting spine graph will remain the Hasse diagram.

6 Experiments

In this section, we demonstrate our algorithm by performing several experiments on synthetic and real stratified space datasets that demonstrate our algorithm's ability to uncover the underlying strata and visualize the output.

6.1 Synthetic Examples

We begin by considering two synthetic examples that appeared in related work in the literature. First, as the authors did in [14] and [10], we sampled 300 points from a spiral and 800 points from a plane in $\mathbb{R}^3$; the spiral and plane intersect transversally. We constructed the adaptive cover tree and scaffolding for threshold choices $\tau \approx 0.1$ (resulting in 185 centers) and $\delta \approx 2.5$. Part (a) of Fig. 9 shows the refined local dimension estimation, and part (b) displays the spine. The red points, labelled as three-dimensional, indicate our method's accurate identification of a non-maximal stratum in the underlying stratified space: namely, the singular stratum where the two one-dimensional spiral pieces intersect the two-dimensional plane. The spine captures how the strata are situated relative to one another.

Our second example, a sphere with an attached curve-fragment, is a dataset that also appeared in [14]. Part (c) of Fig. 9 shows the refined local dimension estimation and part (d) shows the spine. Again our method identifies the important location where two maximal strata of different dimensions meet. We point out that the same spine would have been produced by a point cloud sampled from a 2-plane with a ray sticking out of it. Persistent homology, computed on each node, distinguishes these two cases. In part (d), the spine nodes are decorated with vectors of persistent Betti numbers.

For another example that incorporates topological information, we constructed a point cloud sampled from three lollipop-like structures in $\mathbb{R}^6$ (6000 points in total) belonging to three distinct, mutually orthogonal planes in $\mathbb{R}^6$. Each such structure is made up of a circle and a line segment emanating out from a point on the circle, and the three line segments meet at a single point at each of their bases. See Fig. 10a. Note that our method detects a three-dimensional non-maximal node corresponding to the point of contact of the three line segments, and the two-dimensional non-maximal nodes occurring at the intersections of the circles and line segments. Every node consists of a single persistent connected component. The nodes that have been decorated with vectors of persistent Betti numbers are the nodes with nontrivial one-dimensional homology.

Finally, for our last synthetic example, we consider a 2-plane and a 3-plane intersecting along a one-dimensional subspace in $\mathbb{R}^4$. Our method accurately finds two nodes of dimension two, one three-dimensional node, and one four-dimensional node in the spine. It is evident from the spine, seen in Fig. 10b, that it is impossible to move from one of the two-dimensional maximal nodes to the three-dimensional maximal node without passing through the four-dimensional non-maximal node.

6.2   *Two Experiments on Music Structure Visualization*

We turn to the domain of music analysis to test our methods on real data. Specifically, we address the problem of music structure analysis by representing a song as a point cloud in some feature space and using our methods to automatically group distinct musical segments, such as the chorus and verses, into strata. The problem of automatically recognizing song structure and detecting boundaries between song segments is long standing in the music information retrieval community [21]. Of particular note is recent work in [17] which uses spectral clustering on nearest neighbor graphs of a point cloud in a related feature space to ours, with additional edges added between points adjacent in time. The eigenvectors corresponding to the lower eigenvalues can then be used to encode membership to different song segments, and taking increasing numbers of eigenvectors allows for a multi-scale notion of structure, which is important, since the notion of a song "segment" is otherwise ill-posed [18]. Compared to this technique and other previously published techniques, we are more interested in automatically locating distinct song segments, estimating their local dimensions, and visualizing the results.

Music Feature Space Before applying our pipeline on actual music data, we first provide a high-level overview of our scheme for turning music into point clouds. It is of utmost importance to perform a feature extraction on audio time series, as music data typically consists of very noisy, high-frequency data sampled at 22,050 or 44,100 samples per second which is difficult to analyze in its raw form. The first group of features we use are referred to as "timbral features," which are simple shape features on top of the Short-Time Fourier Transform (STFT) frame meant to summarize relative information across the spectrum [24]. We also use the popular "Mel-Frequency Cepstral Coefficients" (MFCC) [7], which are perceptually motivated coarse descriptors of the shape of the spectral envelope of an STFT window. Absolute pitch information is blurred in the above features, so we also use a feature set known as "chroma" to capture complementary information about notes and chords [1, 13].

All of the features described so far are typically computed over a very small amount of audio, which we call a "window," so that each spectrogram frame is nearly stationary ([24] calls this an "analysis window"). For example, a standard choice is to take this window size to be 2048 samples for audio sampled at 44,100 Hz, or approximately 50 ms. For 5 timbral features, 12 MFCC coefficients, and 12 chroma features, this gives a total of 29 features per window. The features in these windows alone are not appropriate for structure modeling, however, because they are simply too short in time, causing rapid variations from one window to the next and the inability to model higher-level information. To make each feature vector more distinct and the resulting point cloud smoother, we aggregate sets of windows into larger time scales called "blocks" ([24] calls these "texture windows"). In each block, we take the mean and standard deviation of each feature in the windows contained in that block, for a total of 29*2, adding an additional "low energy measure" over the block (as recommended in [24]). This yields a total of 59 feature dimensions per block. Since the features are on different scales, we also normalize each dimension by its standard deviation. In our applications, we typically take the set of all blocks consisting of 150 consecutive windows, spanning between 3 and 7 s per block. This range in block size yields point clouds that are of sufficiently large size to effectively summarize fine audio information and yet are computationally tractable. In sum, we transform each song into a 59 dimensional point cloud, where each point represents perceptually motivated statistics in a small time window.

Spine Graphs for Music Data We demonstrate the results of our pipeline for data organization and visualization purposes on two songs[1]: "Bad" by Michael Jackson (Fig. 11a) and "Lucy in The Sky with Diamonds" by The Beatles (Fig. 12a). Here,

---

[1]Interactive versions of these examples, which allow one to play the song and explore the scaffolding at the same time, are available online at the following location: http://www.ctralie. com/Research/GeometricModels.

the point cloud for "Bad" consists of $\approx$12,000 points (obtained using 3-second windows), and the point cloud for "Lucy" consists of $\approx$4300 points (obtained using 7-second windows). We visualize the point clouds in $\mathbb{R}^3$ by projecting onto the first three principal components (in both cases, approximately 60% of the variance is explained by these components).

Before discussing the output of the algorithm, we wish to first indicate our reasoning behind the choice of eigenthreshold as well as give an idea of the runtime of the method. In the case of the "Bad" example, setting the eigenthreshold $\tau = 1$ ensures that (using the notation of Sect. 3.1) for all $i = 1, \ldots, n$ (with a standard default value of $n = 30$), $||\lambda^i(p) - \lambda^i(q)|| < \epsilon$ with $\epsilon = 0.2$ for any pair of centers p, q being compared via the eigenmetric. Also, setting $\tau = 1$ results in $\approx$ 500 centers in the cover tree, yielding a representative sample of 4–5% of points as centers in the cover tree. The process of extracting features from "Bad" to turn it into a point cloud using 3-second time windows on a 2.3 GHz Intel Core i5 processor had a run time of $\approx$80 s, and it took $\approx$200 s for the cover tree to be built. The local dimension estimation took $\approx$45 s. In contrast, for the "Lucy" example with 7-second time windows and eigenthreshold $\tau \approx 2.5$, the feature extraction process had a run time of $\approx$40 s, the cover tree was built in $\approx$45 s, and the local dimension estimation procedure took less than 3 s.

As evident in Figs. 11b and 12b, the spine graphs visibly capture high-level structural elements of the music for both songs. Overall, our method provides a means of automatically determining the local geometric structure of audio data, and our spine graphs provide musically relevant visuals for acoustic clustering and transitions. In the "Bad" example, an additional distinction is seen between the instrumental only part of the verse and the part with the same vamp (repeating) instrumentals with vocals on top. These song components are acoustically similar but different enough to be clustered into different strata right next to each other in our scheme. Across varying parameter choices, experimental evidence using our method consistently shows that areas of greater musical complexity (e.g., verses, chorus) tend to be characterized by higher-dimensional strata, whereas transitions between distinct parts of the song tend to be represented by one-dimensional strata connecting higher-dimensional segments.

7   Discussion

This paper proposes a novel, fast, and flexible technique for organizing, visualizing, analyzing, and understanding a point cloud that has been sampled from or near a stratified space. We have demonstrated the utility of the method with some experiments on synthetic data, and our real data application in this work was to music structure visualization. Going forward, we plan to build on our methods with the ultimate goal being to match cover songs to original artist versions based on components of their spine graphs, as well as to perform classification by genre.

Our framework is quite general and is well-suited to many kinds of data, not just music. For instance, in the domain of natural language processing, one could

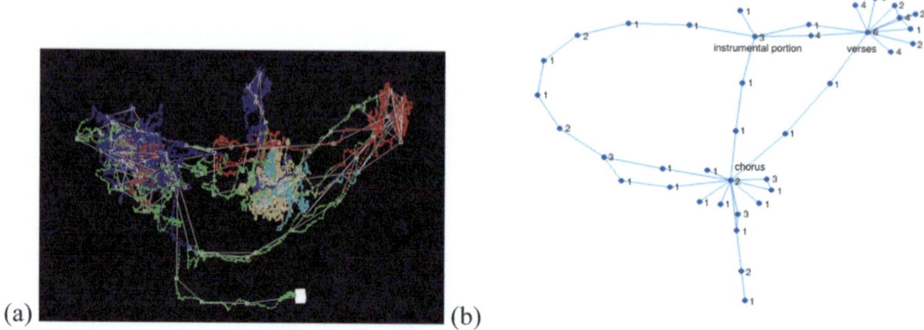

Fig. 11 (**a**) The point cloud corresponding to Michael Jackson's song "Bad," colored by dimension within each node and with the scaffolding edges included. (**b**) The spine graph for "Bad"

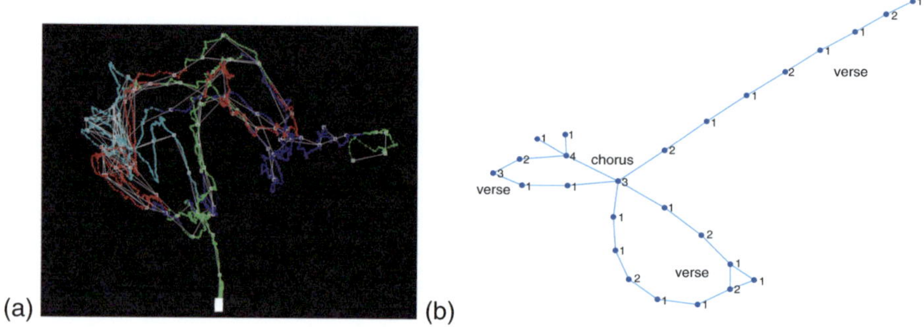

Fig. 12 The point cloud (**a**) and spine graph (**b**) corresponding to The Beatles' song "Lucy in the Sky with Diamonds"

analyze a collection of documents using some metric between them, with the aim of separating out different topics into different strata. With a simple bag of words approach, there is also a natural interpretation of the cover tree centers and eigenvectors within each stratum: The centers capture the words that make up the main topic/conversation, while the eigenvectors capture groups of words that differentiate documents in the same conversation.

We end by outlining two conjectures for future work. First, the scaffolding $\Sigma(\mathbb{X})$ and the spine $S(\mathbb{X})$ are both representations of the point cloud $\mathbb{X}$. We would like to prove a stability result relating these to $\Sigma(\mathbb{Y})$ and/or $S(\mathbb{Y})$ for some point cloud $\mathbb{Y}$ that is near $\mathbb{X}$ in, say, Hausdorff distance. It is possible that a multiscale version of the robust PCA framework introduced in [8], rather than standard PCA, will be useful for analyzing a point cloud sampled with a small amount of noise from a given stratified space. Such a framework could also be useful when dealing with nonuniformly sampled point clouds.

Second and finally, in each of the examples we've shown, the spine of our point cloud $\mathbb{X}$ is exactly the Hasse diagram of the partial ordering on the stratification of the space that $\mathbb{X}$ was sampled from. Compare, for example, the spine in part (c) of Fig. 8 with the space underlying the point cloud in part (a); arrows are not drawn, but imagine that the lower index always points to the higher. Of course, parameters had to be chosen to make this true. We conjecture that this should always work, assuming some appropriate sampling conditions.

References

1. M. Bartsch, G.H. Wakefield, To catch a chorus: using chroma-based representations for audio thumbnailing, in *Applications of Signal Processing to Audio and Acoustics, 2001 IEEE Workshop on the* (IEEE, New York, 2001), pp. 15–18
2. D. Bassu, R. Izmailov, A. McIntosh, L. Ness, D. Shallcross, Centralized multi-scale singular value decomposition for feature construction in LIDAR image classification problems, in *Applied Imagery Pattern Recognition Workshop (AIPR), 2012 IEEE* (2012), pp. 1–6
3. P. Bendich, B. Wang, S. Mukherjee, Local homology transfer and stratification learning, in *Proceedings of the Twenty-Third Annual ACM-SIAM Symposium on Discrete Algorithms* (SIAM, Philadelphia, 2012), pp. 1355–1370
4. P. Bendich, E. Gasparovic, J. Harer, R. Izmailov, L. Ness, Multi-scale local shape analysis and feature selection for machine learning applications, in *Proceedings of International Joint Conference on Neural Networks* (2015)
5. A. Beygelzimer, S. Kakade, J. Langford, Cover trees for nearest neighbor, in *Proceedings of the 23rd International Conference on Machine Learning* (ACM, New York, 2006), pp. 97–104
6. G. Birkhoff, *Lattice Theory* (American Mathematical Society, Providence, 1948)
7. B.P. Bogert, M.J.R. Healy, J.W. Tukey, The quefrency alanysis of time series for echoes: cepstrum, pseudo-autocovariance, cross-cepstrum and saphe cracking, in *Proceedings of the Symposium on Time Series Analysis*, vol. 15 (1963), pp. 209–243
8. E.J. Candes, X. Li, Y. Ma, J. Wright, Robust principal component analysis? J. ACM **58**(3), 11 (2011)
9. F. Chazal, D. Cohen-Steiner, M. Glisse, L. Guibas, S. Oudot, Proximity of persistence modules and their diagrams, in *Proceedings of the 25th Annual Symposium on Computational Geometry, SCG '09* (ACM, New York, 2009), pp. 237–246
10. G. Chen, A.V. Little, M. Maggioni, Multi-resolution geometric analysis for data in high dimensions, in *Excursions in Harmonic Analysis, Volume 1* (Springer, Berlin, 2013), pp. 259–285
11. D. Cohen-Steiner, H. Edelsbrunner, J. Harer, Stability of persistence diagrams. Discret. Comput. Geom. **37**(1), 103–120 (2007)
12. H. Edelsbrunner, J. Harer, *Computational Topology: An Introduction* (American Mathematical Society, Providence, 2010)
13. T. Fujishima, Realtime chord recognition of musical sound: a system using common lisp music, in *Proceedings of ICMC*, vol. 1999 (1999), pp. 464–467
14. G. Haro, G. Randall, G. Sapiro, Translated poisson mixture model for stratification learning. Int. J. Comput. Vis. **80**(3), 358–374 (2008)
15. B. Hughes, S. Weinberger, Surgery and stratified spaces. Surv. Surg. Theory **2**, 311–342 (2000)
16. G. Lerman, Quantifying curvelike structures of measures by using L_2 Jones quantities. Commun. Pure Appl. Math. **56**(9), 1294–1365 (2003)

17. B. McFee, D.P.W. Ellis, Analyzing song structure with spectral clustering, in *ISMIR 2014: Proceedings of the 15th International Conference on Music Information Retrieval: Taipei, 27–31 October 2014* (2014)
18. B. McFee, O. Nieto, J.P. Bello, Hierarchical evaluation of segment boundary detection, in *16th International Society for Music Information Retrieval Conference, ISMIR* (2015)
19. J.R. Munkres, *Elements of Algebraic Topology* (Addison Wesley, Boston, 1993)
20. L. Parsons, E. Haque, H. Liu, Subspace clustering for high dimensional data: a review. SIGKDD Explor. Newsl. **6**(1), 90–105 (2004)
21. J. Paulus, M. Müller, A. Klapuri, State of the art report: audio-based music structure analysis, in *ISMIR* (2010), pp. 625–636
22. F.R.S.K. Pearson, Liii. On lines and planes of closest fit to systems of points in space. Philos. Mag. Ser. 6 **2**(11), 559–572 (1901)
23. B. St. Thomas, L. Lin, L.-H. Lim, S. Mukherjee, Learning subspaces of different dimension (April 2014). arXiv e-prints
24. G. Tzanetakis, P. Cook, Musical genre classification of audio signals. IEEE Trans. Speech Audio Process. **10**(5), 293–302 (2002)

Density of Local Maxima of the Distance Function to a Set of Points in the Plane

Nina Amenta, Erin Wolf Chambers, Tegan Emerson, Rebecca Glover, Katharine Turner, and Shirley Yap

Abstract We show that the set of local maxima of the distance function to a set of points P in the plane, given certain density and packing restrictions, is also dense.

1 Introduction

The *manifold reconstruction* problem takes as input a set of P points in $\mathbb{R}^n$ that are sampled from a smooth manifold S of some lower dimension k. The goal is to return an approximation of S whose quality is some function of the sampling quality; in particular, if the sampling is insufficiently dense or in some cases too irregular, there might not be much we can guarantee about the quality of the reconstruction.

N. Amenta
University of California Davis, Davis, CA, USA
e-mail: amenta@cs.ucdavis.edu

E. W. Chambers (✉)
Saint Louis University, St. Louis, MO, USA
e-mail: echambe5@slu.edu

T. Emerson
Colorado State University, Fort Collins, CO, USA
e-mail: emerson@math.colostate.edu

R. Glover
University of St. Thomas, St Paul, MN, USA
e-mail: rebecca.glover@stthomas.edu

K. Turner
Australian National University, Canberra, Australia
e-mail: katharine.turner@anu.edu.au

S. Yap
California State University East Bay, Hayward, CA, USA
e-mail: shirley.yap@csueastbay.edu

© The Author(s) and the Association for Women in Mathematics 2018
E. W. Chambers et al. (eds.), *Research in Computational Topology*, Association
for Women in Mathematics Series 13, https://doi.org/10.1007/978-3-319-89593-2_7

The well-studied *surface reconstruction* problem handles the case $n = 3$, $k = 2$, but it was quickly discovered [11] that these algorithms and proofs do not transfer immediately to higher dimensions. A series of papers [2–4, 7] handles these difficulties by perturbation of the input points.

Another line of work in both surface reconstruction [5, 8, 9, 12] and, more recently, manifold reconstruction [10] has approached the problem by considering the critical points of the distance function to P; that is, of $d(x) = \min_{p \in P} d_E(x, p)$ where d_E is the usual Euclidean distance function.

These algorithms exploit a beautiful relationship between d and the Voronoi diagram and Delaunay triangulation of P. The Voronoi diagram of P is the polyhedral complex in which every face, of every dimension, consists of the points nearest to a particular subset $S \subseteq P$. Its dual, the Delaunay triangulation, is the complex formed by the convex hulls of every subset S inducing a non-empty face of the Voronoi diagram. The faces of the Voronoi diagram of dimension at most $d - 1$ also coincide with the singularities (non-smooth points) of the distance function d. The minima of d are the points of P themselves. The critical points of d coincide with points at which a face of the Voronoi diagram intersects its dual Delaunay face, and the index of a critical point is the dimension of the Delaunay face. For example, in $\mathbb{R}^3$, the critical points of index 1 are the points at which a Voronoi 2-face intersects its dual Delaunay edge, the critical points of index 2 are the points at which a Voronoi edge intersects its dual Delaunay triangle, and the maxima are vertices of the Voronoi diagram that lie inside their dual Delaunay tetrahedra. These special Delaunay simplices that include critical points have been called *centered* [8] or *anchored* [10] simplices.

In particular, for reconstructing 2-manifolds in dimension n, finding anchored triangles can be an important first step. In the recent work of Khoury and Shewchuk [10], they show that an anchored triangle t must lie near the 2-manifold S and that its tangent space approximates that of S. In general, getting a good estimate of the tangent space of S at a dense set of points usually goes a long way toward estimating S. This leads us to consider the density of anchored triangles.

2 Main Result

In this note, we consider the density of the set of anchored triangles in the Delaunay triangulation of a set P of points in the plane, $\mathbb{R}^2$. We restrict our problem in two other ways. First, we assume that the sampling is roughly uniform. Second, to avoid dealing with boundary conditions, we assume that P is infinite.

Definition 2.1 *We say that point set P is an (ϵ, δ)-sample if no point of the plane is farther than ϵ from a point of P, and no two points of P are closer together than δ, where ϵ and δ are both constants, with $\delta < \epsilon$.*

While this is a simpler situation than surface or manifold reconstruction, we are hopeful that the structural property we establish here will ultimately be useful in more general situations.

Observation 2.2 *In the Delaunay triangulation of an (ϵ, δ)-sample, no edge is shorter than δ or longer than 2ϵ.*

Definition 2.3 (Anchor Point) *Let P be a collection of points in the plane and consider the Delaunay triangulation of P. An anchor point of the triangulation is a Voronoi vertex that is contained in its dual Delaunay triangle, which we call an anchored triangle. Equivalently, an anchor point is a local maximum of the distance function $d : \mathbb{R}^2 \to \mathbb{R}$ such that $d(x) = \min_{p \in P} |x - p|$ for any point $x \in \mathbb{R}^2$.*

We now move on to our main theorem about the density of anchor points in an (ϵ, δ)-sample of the plane.

Theorem 2.4 *Let x be a point and P be an (ϵ, δ)-sample in $\mathbb{R}^2$. There exists an anchor point a in the ball $B(x, R)$, where $R = \frac{4\epsilon^2}{\delta} + 4\epsilon$.*

We prove this bound through the following sequence of lemmas and a construction.

Lemma 2.5 *A triangle T in a planar Delaunay triangulation is anchored if and only if T is acute.*

Proof An anchored triangle T contains its circumcenter. The three vertices of a triangle T divide its circumcircle into three arcs, as in Fig. 1. Due to the inscribed angle theorem, each angle θ in the triangle is opposite an edge subtending an arc of length 2θ on the circumcircle. So if all the angles are acute, all of the corresponding arcs have length at most π, and if the triangle is obtuse, some arc has length greater than π. T fails to contain its circumcenter if and only if the circumcenter is on the side of an edge opposite the triangle's interior, and this edge must thus subtend an arc of length greater than π. Therefore an acute triangle contains its circumcenter, and an obtuse triangle fails to contain its circumcenter.

$\square$

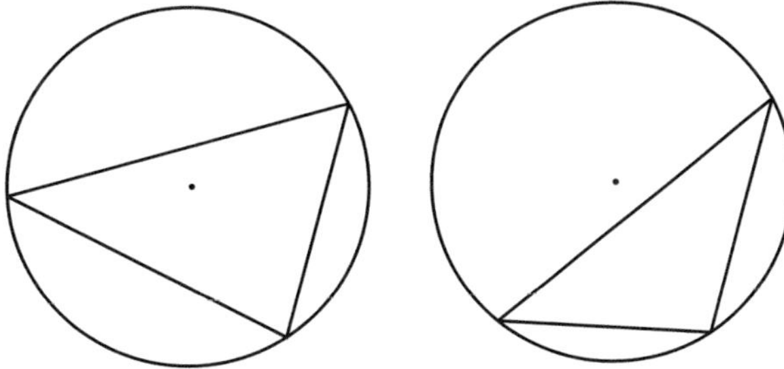

Fig. 1 A triangle strictly contains its circumcenter if and only if it is acute

3 Construction

Let x be an arbitrary point of the plane. Let D be the union of the minimal subset of Delaunay triangles required to cover the Euclidean disk of radius R surrounding x, where R is a constant to be set later. We denote the boundary of D, consisting of a subset of Delaunay edges, by ∂D. Since any triangle in D has to have a point in $B(x, R)$, D is contained in a larger disk of radius $R + 2\epsilon$, by Observation 2.2. Since the distance function d is continuous and ∂D is a compact set, the Extreme Value Theorem tells us that d achieves (at least one) global maximum on ∂D. Assume z is such a maximum, let a be the edge on which z lies, and let T be the triangle contained in D adjacent to a. If T is acute, we are done by Lemma 2.5, since the anchor point in T is within distance $R + 2\epsilon$ of x.

If T is obtuse, then we have two cases: either a is the long side of T or it is one of the shorter sides. We first consider the case where a is one of the shorter sides, and we suppose without loss of generality that b (which may or may not be on ∂D) is the longest side.

Lemma 3.1 *The maximum of the distance function d restricted to any edge a lies at an intersection of some Voronoi edge (possibly but not necessarily dual to a) with a.*

Proof The two endpoints of a are minima of the distance function, and so cannot be maxima. Let y be any point in the interior of a, and consider its position in the Voronoi diagram. If y is in the interior of a Voronoi cell, then the distance function d is the distance function to a single point. Its restriction to edge a is smooth and convex at y, so it might contain a minimum but cannot contain a maximum. Therefore y must lie on a Voronoi edge if it is a maximum of d. $\quad\square$

Lemma 3.2 *Let a be one of the smaller sides of an obtuse Delaunay triangle T, and let z be the maximum of d restricted to a. Then there is a point y, in the interior of T, with $d(y) > d(z)$.*

Proof The maximum z of d on a must lie on some Voronoi edge E, which is the perpendicular bisector of its two closest sample points γ and ρ. If there are no sample points in the interior of the disk D_a with diameter a, then the two closest sample points to z are the endpoints of a and z is the midpoint of a. In this case z is the minimum of d restricted to E, so a small neighborhood around z contains a point y of E with $y \in T$ and $d(y) > d(z)$.

Now suppose that the interior of D_a is not empty. Let C be the circumcircle of T, where T has longest edge $|b| > |a|$. Because T is obtuse, $|b|$ is less than the diameter of C. Further, the semidisk of D_a on the same side of a as b is contained in C, and hence empty of sample points. This shows that γ and ρ are on the opposite side of a as b (see Fig. 2). Since d increases along the perpendicular bisector E of a in either direction away from the sample points, again there is a point $y \in E$ such that $y \in T$ and is within a small neighborhood of z with $d(y) > d(z)$. $\quad\square$

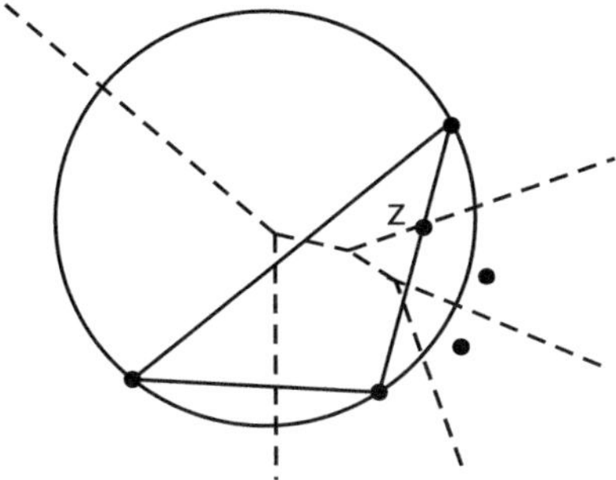

Fig. 2 The Voronoi diagram of the points is shown with dashed lines. The maximum z of the distance function restricted to one of the shorter ends of an obtuse Delaunay triangle lies on a Voronoi edge along which the distance from the points is increasing in the direction of the triangle interior

Corollary 3.3 *Let z be a maximum of the distance function d over ∂D. If z is contained in a Delaunay edge $a \in \partial D$, where a is one of the smaller sides of an obtuse triangle $T \subset D$, then D contains an anchor point.*

Proof Let m be a global maximum of d over the region D. Lemma 3.2 established that, in this case, the maximum value of d in the interior is greater than the maximum value of d on the boundary, so m must lie in the interior of D. $\qquad\square$

We have shown that the only case where a local maximum is not immediate is the case where every global maximum z of d on ∂D lies on the longest edge of an obtuse triangle contained in D, that is, when the edge a containing z is not opposite an acute angle in the triangle bounded by a and in the interior of D. So suppose we have such a global maximum $z \in a$ where a is the longest edge of an obtuse Delaunay triangle T contained in D. We will remove T, including the maximum z, from D. This produces a new, smaller region with a possibly different set of global maxima on its boundary. We repeat this process until we find that we have a maximum z which is opposite an acute angle; we may have to "peel off" many triangles in the process. Note that a triangle removed as part of this peeling process may have one, two, or three sides on ∂D.

Let $\partial D_0 = \partial D$ be the original boundary before we start the peeling process, and ∂D_j be the boundary after removing j triangles. We will prove an upper bound on the distance from ∂D_0 to some global maximum z on some ∂D_j opposite an acute angle. By choosing R larger than this distance, we will guarantee that not all triangles will be peeled away before finding such a maximum z.

Consider a maximum z on ∂D_j; in particular, z is exposed at step j when it appears on ∂D_j but not on ∂D_{j-1}. If z is in the interior of the original region D_0, then it must have been exposed by removing a sequence of triangles $T_1 \ldots T_k$. The last triangle T_k is the one whose removal exposed z. Since T_k was removed, its longest edge a_k was on the boundary of some region $D_{j'}$, $j' < j$. If a_k did not belong to ∂D_0, then it must have been exposed by the removal of some other triangle; this is T_{k-1}, and so on, until we find the first-removed T_1 whose longest edge a_1 does lie on ∂D_0. Note that removing a triangle with three edges on ∂D_j does not expose any new edges, so we can assume that each $T_i \in T_1 \ldots T_k$ had either one or two but not three edges on ∂D_j at the step j at which it is removed.

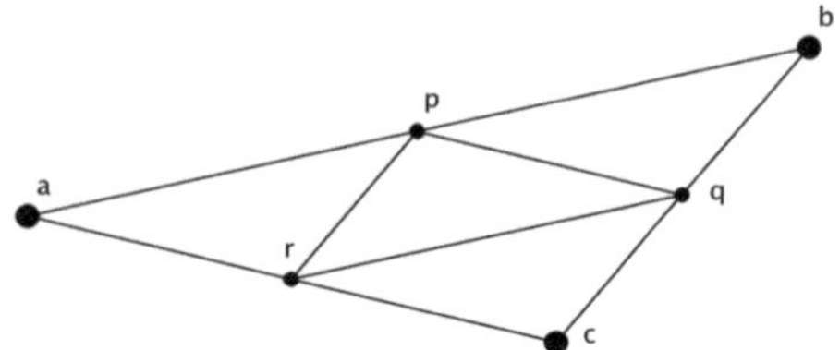

Fig. 3 Note that $\triangle abc$ is similar to $\triangle pqr$, with p, q, and r the midpoints of the sides of the triangle

Also note that each T_i in the sequence of triangles has three edges, its longest edge a_i, the longest edge of the next triangle in the sequence a_{i+1}, and a third edge which we shall call b_{i+1}.

The intuition is that the length of a_i gets shorter at each step, and when a_i has length between $\sqrt{2}\delta$ and δ, it must belong to an acute triangle. But instead of explicitly bounding the number of triangles removed, we can get a better bound by considering the distance from the midpoint of a_1 to the midpoint of a_k.

Lemma 3.4 *Consider a sequence of triangles $T_1, T_2, \ldots, T_k$ constructed as above, and their longest edges $a_1, \ldots a_k$. Let m_i be the midpoint of each a_i. The distance from midpoint m_k to m_1 is at most $4\epsilon^2/\delta$.*

Proof Consider the triangle T_i with edges a_i, a_{i+1}, and b_{i+1}. As shown in Fig. 3, we can argue by similar triangles that $\|m_i - m_{i+1}\| = \dfrac{|b_i|}{2}$. By construction we know that the angle opposite a_i is obtuse. The cosine rule thus tells us that

$$|a_i|^2 = |a_{i+1}|^2 + |b_{i+1}|^2 - 2|a_{i+1}||b_{i+1}|\cos\theta$$

$$> |a_{i+1}|^2 + |b_{i+1}|^2.$$

Further, by algebraic manipulation we have

$$|b_{i+1}|^2 < |a_i|^2 - |a_{i+1}|^2$$

$$= (|a_i| - |a_{i+1}|)(|a_i| + |a_{i+1}|)$$

$$= (|a_i| - |a_{i+1}|)^2 \left(\frac{(|a_i| + |a_{i+1}|)}{(|a_i| - |a_{i+1}|)}\right)$$

$$= (|a_i| - |a_{i+1}|)^2 \left(\frac{(|a_i| + |a_{i+1}|)^2}{|a_i|^2 - |a_{i+1}|^2}\right).$$

We know that $|a_i|^2 - |a_{i+1}|^2 > \delta^2$, and by assumption $(|a_i| + |a_{i+1}|)^2 < (4\epsilon)^2$. This implies that

$$\frac{(|a_i| + |a_{i+1}|)^2}{|a_i|^2 - |a_{i+1}|^2} < \frac{(4\epsilon)^2}{\delta^2}$$

and hence that

$$|b_{i+1}| < \frac{4\epsilon}{\delta}(|a_i| - |a_{i+1}|). \tag{1}$$

We now can use (1) and telescoping sums to bound the sum of the distances between the midpoints as follows:

$$\sum_{i=0}^{k-1} \|m_i - m_{i+1}\| = \sum_{i=1}^{k} \frac{1}{2}|b_{i+1}|$$

$$\leq \sum_{i=0}^{k} \frac{2\epsilon}{\delta}(|a_i| - |a_{i+1}|)$$

$$= \frac{2\epsilon}{\delta}(|a_0| - |a_k|)$$

Just using the fact that $|a_0| < 2\epsilon$, we see that the midpoints of any sequence of obtuse triangles that we peel off can get at most $\frac{4\epsilon^2}{\delta}$ from boundary ∂D_0; using the fact that $|a_k| > \delta$ would improve this bound, but not significantly. $\qquad\square$

Now we complete the proof of the theorem. First, we define a "core" around x, consisting of the minimum subset of the Delaunay triangles that cover a ball of radius 2ϵ around x. The core is chosen large enough that it has to contain at least one complete triangle. We define R, the radius of the disk used to define D, as $\frac{4\epsilon^2}{\delta} + 2\epsilon$ to account for the core and the distance between midpoints. As noted above, D itself extends beyond the disk of radius R but must be contained in a disc of radius $\frac{4\epsilon^2}{\delta} + 4\epsilon$. Thus, any local maximum of d within D has distance at most $\frac{4\epsilon^2}{\delta} + 4\epsilon$ from x.

We use the peeling process to locate a local maximum of d in the interior of D. We perform the peeling process until we end up at a D_j such that the maximum z on ∂D_j is on an edge opposite an acute angle in its corresponding triangle in the interior of D_j. Lemma 3.4 showed that the midpoint of the longest edge of the final obtuse triangle removed is outside of the core, and there is at least one triangle completely contained in the core. So D_j is not empty after the peeling process, and because the peeling process has completed, it must contain a local maximum.

4 Discussion and Open Problems

This result raises a number of interesting questions.

One is whether we can generalize this result in the plane to the situation of surface or manifold reconstruction, for $k = 2$, $n > 2$. This presents a number of difficulties, but some special cases might be easier.

For instance, a triangulated surface T is *self-Delaunay* [1] if the *geodesic* circumcircles on T of all of its triangles are empty of other sample points. Such triangulations exist and can be created by refining an input piecewise-linear surface [6]. It seems likely that we can extend our result fairly directly to self-Delaunay meshes, using the geodesic circles instead of the Delaunay circles we used in the plane.

Another possible easier case was proposed by Khoury and Shewchuk. They showed that it is possible to refine any piecewise-linear surface close enough to S to a piecewise-linear surface M formed by Delaunay triangles, with the property that the dual Voronoi face of each triangle in M intersects M (that is, M is its own *restricted Delaunay triangulation*). It might be possible to extend our result to such a *self-Delaunay* surface M.

Finally, most of the work on surface and manifold reconstruction uses the sampling density condition that no surface point $x \in S$ is farther than ϵ from a sample $p \in P$ but does not require the packing assumption that no two points are closer than δ. It might be possible that our result holds (with a different constant) without the condition on δ.

References

1. A.I. Bobenko, B. Springborn, A discrete Laplace-Beltrami operator for simplicial surfaces. Discret. Comput. Geom. **38**(4), 740–756 (2007)
2. J.-D. Boissonnat, A. Ghosh, Manifold reconstruction using tangential Delaunay complexes. Discret. Comput. Geom. **51**(1), 221–267 (2014)
3. J.-D. Boissonnat, L.J. Guibas, S. Oudot, Manifold reconstruction in arbitrary dimensions using witness complexes. Discret. Comput. Geom. **42**(1), 37–70 (2009)
4. J.-D. Boissonnat, R. Dyer, A. Ghosh, Stability of Delaunay-type structures for manifolds [extended abstract], in *Symposuim on Computational Geometry 2012, SoCG '12, Chapel Hill, 17–20 June 2012* (2012), pp. 229–238
5. T.K. Dey, J. Giesen, E.A. Ramos, B. Sadri, Critical points of the distance to an epsilon-sampling of a surface and flow-complex-based surface reconstruction, in *Proceedings of the 21st ACM Symposium on Computational Geometry, Pisa, 6–8 June 2005* (2005), pp. 218–227
6. R. Dyer, H. Zhang, T. Möller, Delaunay mesh construction, in *Proceedings of the Fifth Eurographics Symposium on Geometry Processing, Barcelona, 4–6 July 2007* (2007), pp. 273–282
7. R. Dyer, G. Vegter, M. Wintraecken, Riemannian simplices and triangulations, in *31st International Symposium on Computational Geometry, SoCG 2015, Eindhoven, 22–25 June 2015* (2015), pp. 255–269
8. H. Edelsbrunner, Surface reconstruction by wrapping finite sets in space. Algorithms Comb. **25**, 379–404 (2003)
9. J. Giesen, M. John, The flow complex: a data structure for geometric modeling. Comput. Geom. **39**(3), 178–190 (2008)
10. M. Khoury, J.R. Shewchuk, Fixed points of the restricted Delaunay triangulation operator, in *32nd International Symposium on Computational Geometry, SoCG 2016, Boston, 14–18 June 2016* (2016), pp. 47:1–47:15
11. S. Oudot, On the topology of the restricted delaunay triangulation and witness complex in higher dimensions. CoRR, abs/0803.1296 (2008)

12. E.A. Ramos, B. Sadri, Geometric and topological guarantees for the WRAP reconstruction algorithm, in *Proceedings of the Eighteenth Annual ACM-SIAM Symposium on Discrete Algorithms, SODA 2007, New Orleans, 7–9 January 2007* (2007), pp. 1086–1095

Mind the Gap: A Study in Global Development Through Persistent Homology

Andrew Banman and Lori Ziegelmeier

Abstract The Gapminder project set out to use statistics to dispel simplistic notions about global development. In the same spirit, we use persistent homology, a technique from computational algebraic topology, to explore the relationship between country development and geography. For each country, four indicators, gross domestic product per capita; average life expectancy; infant mortality; and gross national income per capita, were used to quantify the development. Two analyses were performed. The first considers clusters of the countries based on these indicators, and the second uncovers cycles in the data when combined with geographic border structure. Our analysis is a multi-scale approach that reveals similarities and connections among countries at a variety of levels. We discover localized development patterns that are invisible in standard statistical methods.

1 Introduction

The Gapminder Tools [17] project provides a viewpoint of global development through a statistical lens.[1] The first chart that loads in Gapminder plots each country's gross domestic product (GDP) against the life expectancy of its citizens, see Fig. 1. The project equates GDP per capita with a nation's wealth and life expectancy with its health. Countries are color-coded by their broad geographic region: the Americas, Eurasia, etc. A time lapse animation shows countries transitioning along a common trajectory towards more health and wealth, telling a common story about global development. However, it is not clear what role geography plays in this trend.

---

[1]Free material from gapminder.org.

A. Banman (✉)
University of Minnesota, Minneapolis, MN, USA
e-mail: banma001@umn.edu

L. Ziegelmeier
Department of Mathematics, Statistics, & Computer Science, Macalester College, Saint Paul, MN, USA
e-mail: lziegel1@macalester.edu

© The Author(s) and the Association for Women in Mathematics 2018
E. W. Chambers et al. (eds.), *Research in Computational Topology*, Association for Women in Mathematics Series 13, https://doi.org/10.1007/978-3-319-89593-2_8

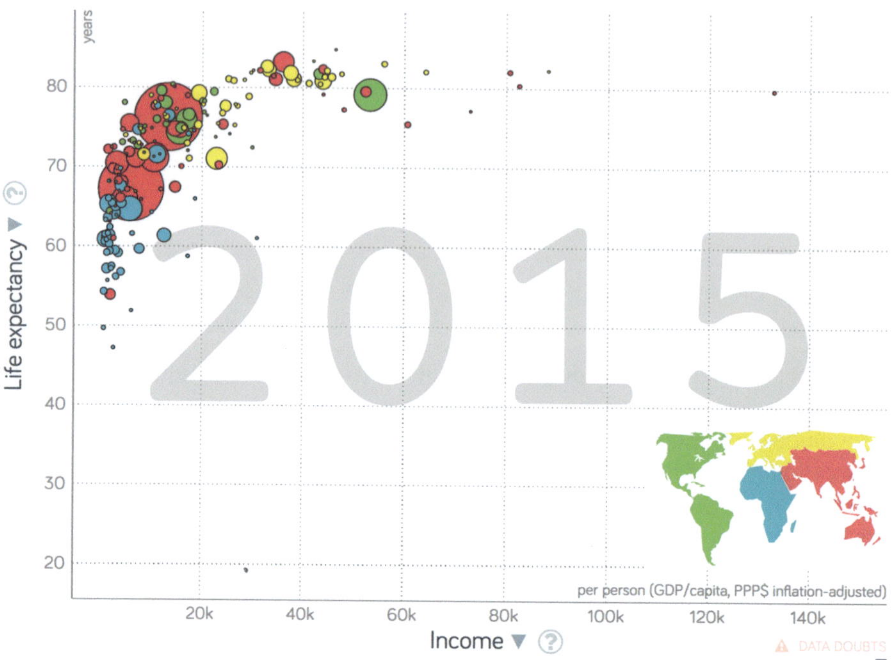

Fig. 1 "Health & Wealth of Nations" chart from Gapminder Tools [10]

While one may say that most African nations lag behind most Eurasian nations, it is difficult to draw any finer conclusions solely from these two statistics, as each region spans a large range of the development statistics. Furthermore, Gapminder's pre-determined regions have been chosen according to a convention rather than from the data. For instance, it splits the African continent into Northern and Sub-Saharan regions, isolates India and a few of its neighbors, and joins Australia with Southern Asian countries. These regions do not necessarily align with regions of differing development.

We seek a quantifiable, fine-grained, and unbiased method to analyze development and geographic trends in this data. Persistent homology [2, 5, 9] gives us tools to uncover the structure of high-dimensional, complicated data, revealing groups (connected components) and cycles (loops) in the data at multiple scales. Persistent homology has been used to understand the topological structure of data arising from applications including computer vision, biological aggregations, brain structure, among many others [3, 4, 11, 14, 15, 21]. In particular, the paper [16] analyzes data related to the recent, so-called, "Brexit" referendum using persistent homology.

We use persistent homology to expand on Gapminder's study of health and wealth statistics. We explore two methods (1) computing the connected components of the indicators of GDP per capita and life expectancy as well as infant mortality and gross national income per capita and (2) adding the underlying geography to the indicators by constructing a weighted graph based on country borders to observe cycles in the data. The structure of the data is uncovered at multiple scales. Our analyses reveal that there are connections among countries at a variety of levels and show subtleties with country similarities and differences, as well as loops formed by countries geographically linked. This provides a more nuanced view than simply the "first" versus "third" world paradigm, a construction that divides the world into discrete sets of developed and undeveloped countries [13].

The remainder of this paper proceeds as follows. Background on the computational approach of persistent homology is discussed in Sect. 2. Section 3 outlines the indicators we use to quantify health and wealth of nations, and our implementation of persistent homology on these indicators. We analyze the results of these computations in Sect. 4. Conclusions and future work are discussed in Sect. 5.

2 Background on Persistent Homology

Persistent homology is a computational approach to topology that encodes a parameterized family of homological features such as connected components, loops, trapped volumes, etc of a topological space. It allows one to answer basic questions about the structure of point clouds at multiple scales. As such, it can uncover the "shape" of data. Broadly, this procedure involves (1) interpreting a point cloud as a noisy sampling of a topological space, (2) creating a global object by forming connections between proximate points based on a scale parameter, (3) determining the topological structure made by these connections, and (4) looking for structures that persist across different scales. For foundational material and overviews of computational homology in the setting of persistence, see [2, 5, 6, 9, 23].

Beginning with a finite set of data points, a nested sequence of simplicial complexes indexed by a parameter ϵ may be created by taking the vertices as the data points and forming a k-simplex whenever $k + 1$ points are pairwise within distance ϵ. This procedure is known as the *Vietoris-Rips* (VR) complex which is often used for its computational tractability [9]. Fixing a field $\mathbb{F}$, one builds a chain complex of vector spaces over $\mathbb{F}$ for each simplicial complex. For each pair $\epsilon_1 < \epsilon_2$, there is a pair of simplicial complexes, S_{ϵ_1} and S_{ϵ_2}, and an inclusion map $j : S_{\epsilon_1} \hookrightarrow S_{\epsilon_2}$. This inclusion map induces a chain map between the associated chain complexes which further induces a linear map between the corresponding kth homology vector spaces. The dimension of the kth homology vector space is known as the kth *Betti number* β_k and corresponds to the number of connected components, loops, trapped volumes, etc. of a simplicial complex for $k = 0, 1, 2, \ldots$, respectively.

The kth barcode is a way of presenting Betti numbers across multiple scales ϵ [9]. From the barcode, one can visualize the number of independent homology classes that persist across a given filtration interval $[\epsilon_b, \epsilon_d]$ as a function of the scale ϵ. See the top row of Fig. 3a for an example β_0 barcode and the bottom row of Fig. 3a for an example β_1 barcode. Each horizontal bar begins at the scale where a topological feature first appears ("is born") and ends at the scale where the feature no longer remains ("dies"). The kth Betti number at any given parameter value ϵ is the number of bars that intersect the vertical line through ϵ. For β_0 in our setting, there will be a distinct bar for each data point at small values of ϵ, as the simplicial complex S_ϵ consists only of isolated points. At large values of ϵ, only one bar remains as all data will eventually connect into a single component.

The idea of persistence is to not only consider the homology for a single specified choice of parameter ϵ but rather, track topological features through a range of parameters. Those which persist over a large range of values are considered signals of underlying topology, while the short lived features are taken to be noise inherent in approximating a topological space with a finite sample [2].

3   Methods

There are many ways to quantify the health and wealth of nations. We study four development indicators using the freely-available data from [17]: gross domestic product (GDP) per capita,[2] life expectancy,[3] rate of infant mortality,[4] and gross national income (GNI) per capita.[5] These indicators were chosen because (1) we believe them to be broad indicators of health and wealth, and (2) recent data is available for a large set of countries in each indicator.

We consider this data in two sets: what we will call the four-dimensional ($\mathbb{R}^4$) data comprising all four indicators and the two-dimensional ($\mathbb{R}^2$) data comprising only GDP/capita and life expectancy. The raw $\mathbb{R}^2$ data—before scaling as discussed below—generates the Gapminder chart, see Fig. 1, allowing a comparison of our results to the chart.

The frequency of reporting and currency of statistics can vary dramatically by country so any result necessarily carries the "according to available data" qualifier. We construct our data sets by taking the most recent value for each indicator corresponding to a country.[6] Countries with no available data for one or more indicators in this time frame are excluded from the data set. This yields data comprising 194 countries in the $\mathbb{R}^2$ set and 179 countries in $\mathbb{R}^4$. See Table 1 for statistics such as the maximum, minimum, median, mean, and standard deviation for the raw data of the indicators.

We consider the relative health and wealth of countries, and the presence of extreme outliers in GDP obscures this relationship. Rather than exclude these

---

[2]Gross Domestic Product per capita by Purchasing Power Parities (in international dollars, fixed 2011 prices). The inflation and differences in the cost of living between countries has been taken into account [19].

[3]The average number of years a newborn child would live if current mortality patterns were to stay the same [20].

[4]The probability that a child born in a specific year will die before reaching the age of one, if subject to current age-specific mortality rates. Expressed as a rate per 1000 live births [10].

[5]Gross national income converted to international dollars using purchasing power parity rates [22].

[6]Most data comes from years 2015, 2016, with others as early as 2005. See Table 8 in Appendix 1.

Table 1 Statistics of each indicator: GDP per capita (GDP), Life Expectancy (LE), Infant Mortality rate (IM), and GNI per capita (GNI)

Indicator	Max	Min	Median	Mean	Stand dev	Scaled mean
GDP	148,374	599	11,903	18,972	21,523	−0.476
LE	84.8	48.86	74.5	72.56	7.74	0.296
IM	96	1.5	23.89	15	21.9	0.528
GNI	87,030	350	8360	13,596	15,399	−0.431

The first five statistics correspond to the raw data; the last corresponds to the attenuated and scaled data. Naturally, high GDP, GNI, and life expectancy are favorable, whereas high infant mortality rate is unfavorable

countries outright, we modulate their values to two standard deviations from the mean. Alternatively, we could have taken the logarithm of GDP to bring the outliers closer to the bulk. However, this option has the undesirable consequence of exaggerating the distance between countries with very low GDP and understating the distance between higher GDP countries. For our purposes, it made more sense to collect the richest countries into one group at the extreme of the spectrum and likewise for the poorest. The same attenuation was done for the GNI per capita indicator.

Each indicator is then re-scaled to $[-1, 1]$. The range $[-1, 1]$ was chosen to give a normative representation of each indicator, in which -1 is least favorable and 1 is most favorable, e.g. the country with lowest life expectancy has -1 for that dimension and the country with lowest infant mortality has 1 in that dimension. Note that this does not imply zero is the average value for any indicator. There are many more relatively low GDP countries, even after attenuating outliers, see Table 1. This scaling is required to ensure each indicator carries equal weight in the persistent homology calculations. Otherwise GDP/capita and GNI/capita would completely obscure any features in life expectancy and infant mortality rate because they are orders of magnitude larger in conventional units.

For our calculations, we use the TDA library in R [8]. This library provides an API to create a filtered simplicial complex upon which to calculate the persistent homology. The final result of the computation is a list of persistence intervals $[\epsilon_b, \epsilon_d]$, neatly displayed in a homology barcode, where each interval indicates a homological feature that is born at ϵ_b and dies at ϵ_d. In this section, we outline our procedure for computing persistent homology of our data. In the next, we analyze the results.

For our first experiment, we interpret each set of countries as a point cloud with each indicator value as a dimension. We then apply the Euclidean metric to define the distance between two countries x and y over a set of indicators I:

$$d_I : \mathbb{R}^{|I|} \to \mathbb{R}$$

$$d_I(x, y) = \sqrt{\sum_{i \in I}(x_i - y_i)^2}$$

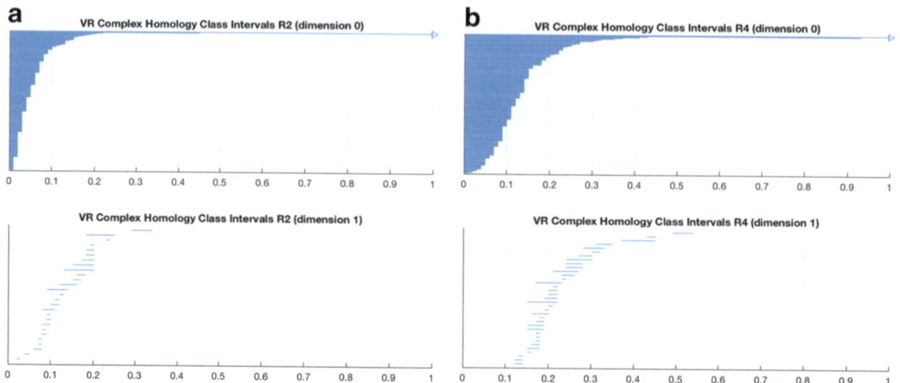

Fig. 2 Zero-order (top) and first-order (bottom) persistent homology barcodes of the VR complex stream over point cloud in $\mathbb{R}^{|I|}$: (**a**) $I = \{\text{GDP}, \text{LifeExp}\}$ and (**b**) $I = \{\text{GDP}, \text{LifeExp}, \text{InfMort}, \text{GNI}\}$

We use TDA to construct a stream of VR complexes from these point clouds over a range of filtration values $\epsilon \in [0, 1.0]$. Figure 2 shows the zero-order and first-order barcodes of the VR streams for the two sets of indicators ($\mathbb{R}^2$ on the left and $\mathbb{R}^4$ on the right).

For the second experiment, we add the geographic structure to the data by constructing a weighted graph over the countries and their borders. From country border data [18], we define an adjacency matrix A

$$A_{(i,j)} = \begin{cases} 1 \text{ if countries } i, j \text{ share a border,} \\ 0 \text{ if countries } i, j \text{ do not share a border} \end{cases}$$

from which we arrive at the distance matrix D for a set of indicators I,

$$D_{I(i,j)} = \begin{cases} d_I(i, j) \text{ if } A_{i,j} = 1, \\ \infty \qquad \text{ if } A_{i,j} = 0 \end{cases}$$

where, for practicality, infinity is set to be a number larger than the maximum filtration value. This maximum is chosen to be large enough to display the entire set of intervals. We then compute the persistent homology of the explicit metric space defined by D_I.[7] The zero-order and first-order persistent homology barcodes for the weighted graphs over $\mathbb{R}^2$ data and $\mathbb{R}^4$ data are shown in Fig. 3. In this framework incorporating geographic structure, our focus is on the first-order features.

---

[7] It has been observed that, for the VR complex, the metric in question need not actually be a metric as it is not a requirement to satisfy the triangle inequality [1]. The construction described here is also known as a weighted rank clique complex. For example, see [16].

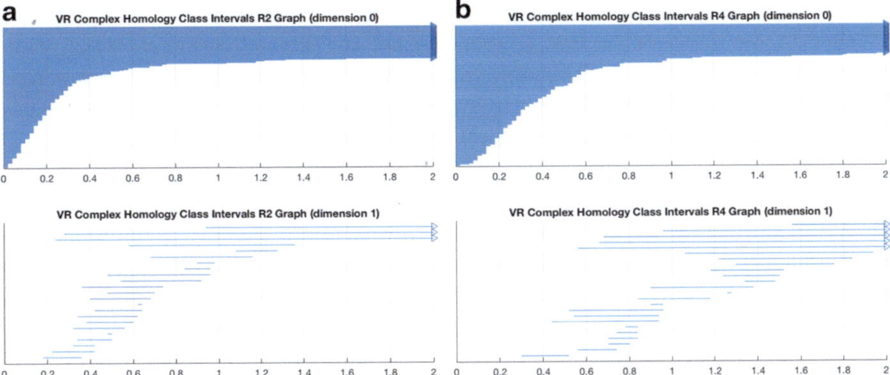

Fig. 3 Zero-order (top) and first-order (bottom) persistent homology barcodes of the VR complex stream over country border graph with distance d_I as the edge weight. (**a**) $I = \{\text{GDP, LifeExp}\}$ and (**b**) $I = \{\text{GDP, LifeExp, InfMort, GNI}\}$

Generally, longer intervals are construed to represent more significant homology classes while short intervals are noise in the data. Statistically significant intervals can be quantitatively determined by the methods presented in [7]. However, we shall see that even relatively short intervals in the first-order barcode reveal interesting patterns in the development indicators. On the other hand, intervals in Fig. 3 that persist through the full range of the filtration are less interesting to us as they relate to the inherent border graph structure. These "infinite" intervals in the dimension-0 barcode indicate island nations that share no borders with other countries. Since their distance to all other countries is infinite, they remain distinct components in the VR complex. The infinite intervals in the dimension-1 barcodes indicate homology classes inherent to the country border graph. The three infinite intervals in Fig. 3a identify the Black, Caspian, and Mediterranean seas. Figure 3b has two additional intervals that exist because two countries (South Sudan and Zimbabwe) were dropped from the data set as not all four indicators were present, creating holes in the graph not unlike an inland sea. That these features are identified is a good sanity check for the method.

4 Parsing the Barcodes

4.1 Clustering of Development Groups

Zero-order persistent homology can be viewed as a clustering algorithm, where the connected components of a simplicial complex represent clusters in the data. In fact, these components are equivalent to the clusters of the hierarchical method of single-linkage clustering. In Fig. 1, we see a clustering chosen by Gapminder. In this section, we describe the clusters found using zero-order persistent homology present in the barcode of Fig. 2, focusing on the first experiment which only relies on distance between indicators and does not incorporate the country border information. In Appendix 2, we present clusters selected by the classic K-means algorithm. Each of these methods results in different clusters. However, we observe that viewing clusters at multiple scales and adding more indicators provides additional insight into relations among countries in terms of health and wealth.

We examine the clusters found using dimension-0 persistent homology by extracting the elements in each component of the simplicial complex for a particular filtration value, see the top row of Fig. 2. One may imagine drawing a vertical slice through the dimension-0 barcode at a given ϵ to select the components. We then extract the list of countries comprising each component using a union-find algorithm. The Betti number can be viewed as a function of the filtration value, $\beta_k(\epsilon)$. When $\epsilon = 0$, each country is an isolated point, and hence, $\beta_0(0) = 194$ for the $\mathbb{R}^2$ data and $\beta_0(0) = 179$ for the $\mathbb{R}^4$ data. All countries in the point cloud eventually merge into a single connected component. This occurs at approximately $\epsilon = 0.45$ for the $\mathbb{R}^2$ data and $\epsilon = 0.92$ for the $\mathbb{R}^4$ data, as seen in the barcodes where only one bar remains.

Figures 4 and 5 display the six[8] components that contain the largest number of countries in the cluster at a variety of filtration scales for the $\mathbb{R}^2$ and $\mathbb{R}^4$ data, respectively. We further inspect these components in detail below.

First, we consider the large-scale structure of the data. For the $\mathbb{R}^2$ point cloud there are 170 countries in a single connected component at $\epsilon = 0.14$, eight countries in the next largest, and the remaining countries isolated in small components. We may say this large cluster is the dominant feature of the data. The $\mathbb{R}^4$ point cloud shows the same behavior. Figures 4 and 5 show how quickly this dominant component grows at early filtration values. At no point do we observe two dominant clusters capturing a combined majority of countries.

Thus, the dimension-0 clustering shows that countries of the world may not be neatly divided into "first world" and "third world" categories with this method.[9] The vast majority of countries are statistically quite similar to another country, which itself is similar to some other country, and so on. The result is a gradient in health and wealth statistics, rather than a discrete grouping. This is easily visualized in the Gapminder chart Fig. 1. One sees the countries of the world arrayed along a gradient from poorer countries with less longevity to richer, longer living countries. Persistent homology clustering captures this gradient as the resulting clusters from this method connect points to their nearest neighbors which each connect to their nearest neighbors and so on. This may result in long clusters whose elements at the ends of a cluster may be quite different from one another but are connected through their neighbors.

---

[8]The choice of six is to coincide with the six clusters in the Gapminder project, see Fig. 1.

[9]The clustering presented in Appendix 2 results in different clusters, which more closely align with this simplistic notion.

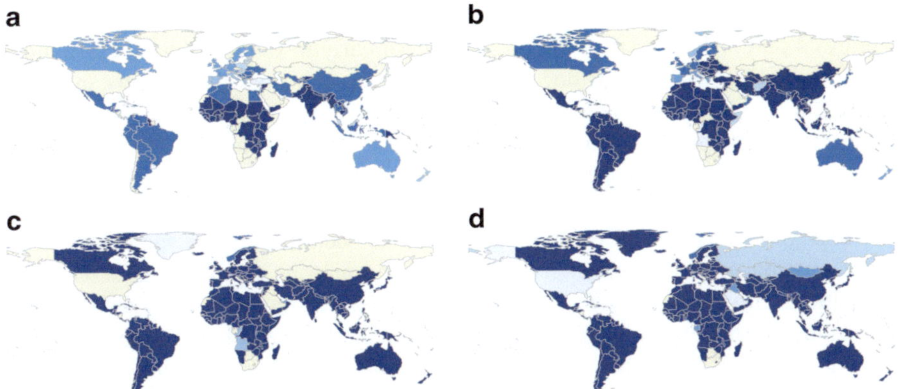

Fig. 4 World map depicting clusters found using dimension-0 persistent homology of the VR complex of the $\mathbb{R}^2$ data at various filtrations. The six largest connected components are displayed in shades of blue (darker indicates larger) while other countries not in these clusters are displayed in yellow. (**a**) $\epsilon = 0.08$ with six largest components consisting of 54, 52, 14, 10, 8 countries among 41 total distinct clusters; (**b**) $\epsilon = 0.10$ with six largest components consisting of 132, 18, 10, 6, 2 countries among 25 total distinct clusters; (**c**) $\epsilon = 0.12$ with six largest components consisting of 164, 6, 2, 2, 2 countries among 19 total distinct clusters; (**d**) $\epsilon = 0.14$ with six largest components consisting of 170, 8, 3, 2, 2 countries among 13 total distinct clusters

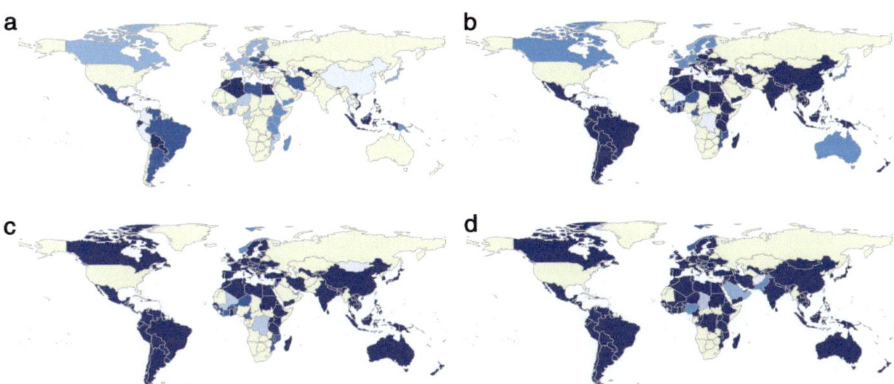

Fig. 5 World map depicting clusters found using dimension-0 persistent homology of the VR complex of the $\mathbb{R}^4$ data at various filtrations. The six largest connected components are displayed in shades of blue (darker indicates larger) while other countries not in these clusters are displayed in yellow. (**a**) $\epsilon = 0.14$ with six largest components consisting of 26, 18, 13, 11, 8 countries among 76 total distinct clusters; (**b**) $\epsilon = 0.16$ with six largest components consisting of 101, 13, 12, 7, 3 countries among 44 total distinct clusters; (**c**) $\epsilon = 0.18$ with six largest components consisting of 115, 13, 7, 3, 2 countries among 40 total distinct clusters; (**d**) $\epsilon = 0.20$ with six largest components consisting of 133, 7, 5, 3, 2 countries among 33 total distinct clusters

We also examine the small-scale structure by looking at smaller ϵ cross-sections. Figures 4 and 5 show a sampling of clusters for early filtration values, before most countries are joined up into one dominant cluster. Consider the clusters in $\mathbb{R}^2$ at $\epsilon = 0.08$, shown in Fig. 4a and detailed in Table 2. While most countries fall into connected components of one to four countries, there are six larger components that capture 138 countries. Because these clusters only exist at a small scale, the countries in each cluster must be quite close in the data. Hence, we may conceive of these groups as sets of very similar countries according to the indicators. This clustering makes a distinction between groups of countries with varying GDP/capita and similar life expectancy. Observe clusters 2–4 have similar life expectancy but a wide range of increasing GDP. Likewise, clusters 5, 6 have almost the same LE but a 0.4 gap in GDP. From this result we may conclude there is nuance in development among poor countries that may be obfuscated by the "third-world" identifier.

Table 2 Countries comprising the largest connected components in the VR complex at filtration $\epsilon = 0.08$ over $\mathbb{R}^2$ and the corresponding means of scaled indicators, GDP/capita (GDP) and life expectancy (LE), for each cluster

Countries (ISO2)	GDP	LE
Bangladesh, Kyrgyzstan, Cambodia, Mauritania, Micronesia Fed. Sts., Nepal, Syria, Gambia, Comoros, Myanmar, Sudan, Sao Tome and Principe, India, Laos, Marshall Islands, Guyana, Pakistan, Ghana, Nigeria, Yemen Rep., Djibouti, Kenya, Senegal, Tanzania, Vanuatu, Haiti, Liberia, Madagascar, Solomon Islands, Ethiopia, Rwanda, Benin, Kiribati, Burkina Faso, Burundi, Congo Dem. Rep., Niger, Papua New Guinea, Togo, Uganda, Zimbabwe, Eritrea, Mali, Malawi, Guinea, Cote d'Ivoire, Cameroon, Sierra Leone, Mozambique, Chad, Zambia, South Sudan, Guinea-Bissau, Fiji	−0.93	−0.15
Albania, Bosnia and Herzegovina, Colombia, Jordan, Sri Lanka, Tunisia, Peru, Macedonia FYR, Barbados, China, Dominican Rep., Algeria, Ecuador, Montenegro, Serbia, Thailand, Bulgaria, Brazil, Iran, Venezuela, Mauritius, Mexico, Romania, Argentina, Saint Lucia, Armenia, Jamaica, Paraguay, El Salvador, Morocco, Vietnam, Bolivia, Bhutan, Cape Verde, Georgia, Guatemala, Honduras, Moldova, Samoa, Belize, Ukraine, Indonesia, Philippines, Saint Vincent and the Grenadines, Egypt, Grenada, Tonga, Uzbekistan, Tajikistan, Korea Dem. Rep., Timor-Leste, Palestine	−0.69	0.44
Antigua and Barbuda, Croatia, Uruguay, Cuba, Panama, Turkey, Lebanon	−0.37	0.63
Estonia, Poland, Slovak Republic, Hungary, Latvia, Malaysia, Lithuania, Seychelles	−0.19	0.53
Cyprus, Malta, Slovenia, Israel, Spain, Italy, Korea Rep., New Zealand, Portugal, Greece	−0.02	0.83
Austria, Australia, Canada, Germany, Denmark, Netherlands, Sweden, Belgium, Taiwan, Finland, France, United Kingdom, Bahrain, Ireland	0.38	0.80

Clusters are listed in ascending GDP order, for clarity in comparison

One advantage of persistent homology as a clustering algorithm is the total lack of bias in the origination of each cluster. Further, a smaller filtration ϵ yields a finer clustering, whereas a relatively large ϵ reveals a coarser structure of the data allowing for a multi-scale analysis. However, the algorithm is highly sensitive to "bridge" structures that connect one cluster to another, destroying distinct components. A bridge in our data might be a relatively poor country with high longevity connecting to a relatively wealthy country with similar longevity, thus joining a cluster of poor countries with a cluster of wealthier countries.

4.2 *Local Development Patterns*

First-order homology classes represent cycles in the data, often visualized as loops around a hole in the point cloud or graph. The dimension-1 barcode intervals tell us over what range of filtration values these cycles exist. Software provides a list of generating simplicies for each homology class, which we parse as a list of generating countries. These generators tell us where in the world the cycle exists.[10]

Our main focus for dimension-1 homology is on the weighted border graphs (Fig. 3). The most discernible intervals are those persisting through the full range of the filtration. As discussed in Sect. 3, these infinite loops describe the topology of the border graph itself. More interesting are the cycles not inherent to the graph structure. These cycles exist because of a pattern of similarity between country neighbors in the indicator data. We map out the generating countries of the six longest-persisting (non-infinite) cycles from the $\mathbb{R}^2$ data in Fig. 6 and further

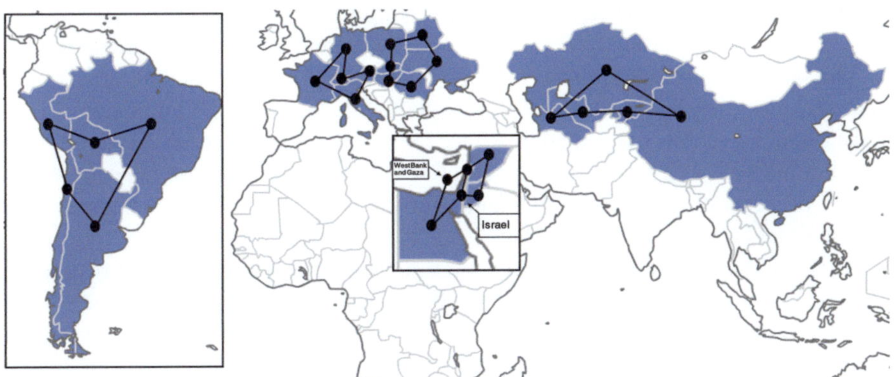

Fig. 6 Map of six cycles in the country border graph with distance d_I, where $I = \{\text{GDP, LifeExp}\}$, as the edge weight. Software-provided generators for each cycle are shown, and the involved countries are shaded

---

[10]The generating countries are not guaranteed to be minimal in a geometric sense; they can make up any loop through the connected component that contains the homology class. One can find the minimal loop by examining the weight of its internal edges.

Table 3 Generating countries of the South America cycle in the $\mathbb{R}^2$-weighted graph from the dimension-1 barcode interval [0.34, 0.62) in Fig. 3a

Country	GDP	LE
Chile	−0.29	0.71
Peru	−0.63	0.72
Bolivia	−0.81	0.37
Brazil	−0.52	0.43
Argentina	−0.45	0.55

Table 4 Generating countries of the North Africa cycle in the $\mathbb{R}^2$-weighted graph from the dimension-1 barcode interval [0.85, 0.97) in Fig. 3a

Country	GDP	LE
Libya	−0.46	0.36
Niger	−0.99	−0.31
Mali	−0.96	−0.36
Mauritania	−0.89	0.17
Algeria	−0.58	0.54

highlight the countries generating two of these cycles in Tables 3 and 4. The cycles are distinguished by a periodic pattern in the data, in which a "maximal" country has the greatest value in one or more indicators; a "minimal" country has the least in these indicators; and the connected countries have intermediary values.

For example, consider the South American cycle in Table 3. A generator of this cycle has Chile with the highest GDP per capita and life expectancy at $(-0.29, 0.71)$ and Bolivia with the lowest at $(-0.81, 0.37)$. Each indicator decreases as you follow the cycle from Chile to Bolivia,[11] and increases on the way back around. The same result holds for the North Africa cycle displayed in Table 4, where Libya is maximal and Mali is minimal. Thus, persistent homology has identified a set of nearby countries that conform to a cycle in both health and wealth statistics. In other words, we've identified a maximal and a minimal country, local to a connected region, where neighboring countries exist on a gradient between the two poles.

The same patterns are found among the $\mathbb{R}^4$ cycles. Almost all of the cycles found in the $\mathbb{R}^4$ case are present in the $\mathbb{R}^2$ case, albeit with different filtration values for ϵ_b and ϵ_d. Some cycles not repeated in $\mathbb{R}^4$ are those involving countries that don't exist in the smaller $\mathbb{R}^4$ data set, due to missing data in the added indicators. In some cases the set of generating countries changed, but the core members—i.e. the maximal and minimal countries—are the same.

Additional cycles in the $\mathbb{R}^4$-weighted graph were created or extended by adding the new indicators, infant mortality and GNI/capita (PPP). Tables 5 and 6 show the adjusted indicator values for two such cycles. The first shows periodic patterns in each of the four indicators from maximal Libya to minimal Chad. This cycle length grew from 0.50 in the $\mathbb{R}^2$ case to 0.90 in $\mathbb{R}^4$. The second cycle is an example of a cycle that wasn't already present in the $\mathbb{R}^2$-weighted graph. The four countries are very close in GDP/capita and life expectancy, but there is strong periodic behavior in infant mortality from maximal Senegal to minimal Mali.

---

[11]There is a slight deviation from monotonic decrease in life expectancy at Peru. These deviations are not uncommon, but do not detract from the maximal-minimal pattern we observe.

Table 5 Cycle from Libya to Chad found in the country border graph with weight d_I, where I={GDP/capita (GDP), life expectancy (LE), infant mortality (IM), GNI/capita (GNI)}

Country	GDP	LE	IM	GNI
Libya	−0.46	0.36	0.79	−0.28
Sudan	−0.89	0.05	0.02	−0.93
Chad	−0.95	−0.49	−0.77	−0.96
Niger	−0.99	−0.31	−0.18	−0.98

Parsed from the interval persisting over [1.10, 1.96) in Fig. 2b

Table 6 Cycle from Senegal to Mali found in the country border graph with weight d_I, where I={GDP/capita (GDP), life expectancy (LE), infant mortality (IM), GNI/capita (GNI)}

Country	GDP	LE	IM	GNI
Mauritania	−0.89	0.17	−0.35	−0.91
Senegal	−0.95	−0.07	0.15	−0.93
Guinea	−0.98	−0.40	−0.26	−0.97
Mali	−0.96	−0.36	−0.54	−0.97

Parsed from the interval persisting over [0.57, 0.75) in Fig. 2b

The relative scales when cycles are born and die reveal how similar the member countries are to one another; the sooner in the filtration they appear, the more similar we can say they are. This follows from the observation that the beginning of an interval is equivalent to the maximum weight of the cycle's edges. Then, we expect countries in later cycles to be further apart, i.e. less similar in the data, than countries in earlier cycles. The South American cycle in Table 3 is a good example of an early cycle showing fine differences in development among countries that are quite similar. Their similarity is on display in the indicator cluster maps that show them often placed in the same cluster (Figs. 5 and 8b).

Note that the death filtration ϵ_d of each cycle coincides with the birth of the simplex that closes the loop. For example, consider the Northern Africa interval, [0.85, 0.97), in Table 4. There are two possible internal edges: Niger $\to$ Algeria and Mali $\to$ Algeria, that come into existence at $\epsilon = 0.94$ and $\epsilon = 0.97$, respectively. All five countries make up the cycle over [0.85, 0.94), but it shrinks at $\epsilon = 0.94$ when N $\to$ A forms. At this point, Libya is cut off from the cycle, which persists with the four other countries until the M $\to$ A simplex closes the loop at $\epsilon_d = 0.97$.

The birth of the closing simplex, i.e. the death of the cycle as a whole, indicates the overall development disparity between countries in the cycle. Compare the cycle from Chile to Bolivia with the cycle from Israel to Syria; the former interval has closing simplex at distance d_I(Bolivia, Brazil) $= 0.63$, while the latter has it's close at distance d_I(Israel, Syria) $= 1.16$. The greater distance between countries in the Israel cycle translates into greater developmental disparity in that region than in South America.

Even relatively short intervals can identify these local development features. The cycle of Afghanistan to Iran (see Table 7) is relatively short with length 0.21 but has one of the largest closing distances, $d_I = 1.30$. This gap in development may also be visualized in Figs. 5b and 8b in that the two countries occupy different development clusters. Hence, cycles may identify the boundaries of clusters found by persistent homology and other methods such as K-means.

Table 7 Countries composing generating cycles and the corresponding birth and death values representing the dimension-1 homology classes of the VR complex stream built over country border graph with weights d_I where I={GDP/capita (GDP), life expectancy (LE), infant mortality (IM), GNI/capita (GNI)}

Birth	Death	Generating countries
0.31	0.52	Hungary, Romania, Croatia, Montenegro, Serbia
0.46	0.94	Chile, Peru, Brazil, Argentina
0.53	0.96	Romania, Ukraine, Belarus, Poland, Hungary, Slovak Republic
0.54	0.94	Austria, Italy, Switzerland, Germany, France
0.56	0.75	Mali, Mauritania, Senegal, Guinea
0.71	0.85	Congo Dem. Rep., Zambia, Tanzania, Burundi
0.71	0.81	Kazakhstan, Turkmenistan, China, Kyrgyzstan, Uzbekistan
0.75	0.85	China, Nepal, Bhutan, India
0.78	0.85	Congo Dem. Rep., Uganda, Burundi, Tanzania
0.84	1.18	Czech Rep., Germany, Austria, Slovenia, Hungary, Slovak Republic
0.90	1.38	Congo Dem. Rep., Congo Rep., Central African Rep., Cameroon
0.91	0.96	Syria, Turkey, Iraq, Iran
1.06	1.95	Algeria, Mauritania, Sudan, Chad, Egypt, Niger, Mali, Libya
1.18	1.52	Israel, Jordan, Lebanon, Syria
1.22	1.85	Afghanistan, Turkmenistan, China, India, Tajikistan, Pakistan, Uzbekistan
1.24	1.51	Algeria, Niger, Mauritania, Mali
1.26	1.28	Afghanistan, Tajikistan, Turkmenistan, Uzbekistan
1.30	1.77	Iran, Pakistan, Afghanistan, Turkmenistan
1.34	1.49	Egypt, Israel, Jordan, Palestine

Cycles are listed in ascending order of interval birth

5 Conclusions and Further Work

Our results show that simplistic notions of country development such as the paradigm of classifying countries as "first" and "third" world masks differences in development among countries. We find that persistent homology as a clustering algorithm does not identify two distinct clusters. PH, on the other hand, discovers fine-grained differences between countries in these categories that are hidden by Gapminder's visualization and K-means clustering. When comparing groups of countries we find that wealth data varies widely between groups with similar values in the health data. While countries may be below average in wealth indicators, like GDP and GNI per capita, they may have quite favorable health indicators, especially life expectancy. Bimodal paradigms conceal this fact.

We also find geographically localized patterns that are invisible in the data consisting of just indicators in $\mathbb{R}^2$ and $\mathbb{R}^4$ by adding country border information into our analyses. First-order PH identifies cycles of development statistics among neighboring countries. In particular, these cycles identify regions of developmental disparity, be it a subtle difference between countries as in the Brazil-Bolivia cycle or a larger gap as in the Israel-Syria cycle. Gapminder's pre-determined regions obfuscate these features as a country's membership to a region implies similarity with the other members. These cycles tell a story about development in a region that would otherwise be masked by other methods.

There are many avenues for further work with our methods. We only consider four indicators, but one may replace, add, or remove indicators as they wish to conduct studies on development or other topics. Our method allows any number of variables to be encoded either as points in a higher dimensional space or as weights in the country border graph. Gapminder hosts a bounty of indicators that may be compared in myriad combinations.

Additionally, one may conduct a longitudinal study of persistent homology using development statistics. We use only the most recent data in our study, but there are decades worth of statistics available. Such a study would need to solve the problem of missing data. Incorporating longitudinal data would make the Betti numbers a function of both time and the filtration scale. One approach to visualize such data is the CROCKER plot discussed in [21], but more techniques may become available as the theory of multi-parameter persistence is an ongoing, active area of research.

Appendix 1

See Table 8.

Appendix 2

The K-means clustering algorithm is an ubiquitous vector quantization method in data mining [12]. The algorithm produces a partitioning of a point cloud of data into K clusters by identifying a set of centers (or prototypes) for each of the clusters, assigning each data point to the cluster with the closest center, calculating the mean of all points in each cluster, and then updating the center of each cluster to be equal to the mean. The process is iterated until all data points are quantized to an appropriate degree of accuracy. A typical implementation of the algorithm is to randomly initialize starting centers and then to iterate for a large number of trials with the goal of minimizing the within-cluster sum of squares error.

We implement the K-means clustering algorithm on our datasets. We mention a couple drawbacks of this method as opposed to persistent homology: (1) The K-means algorithm requires a fixed number of clusters while persistent homology allows for clustering at multiple scales. Therefore, when an appropriate number of clusters is not known *a priori*, specifying a set number of clusters may introduce bias. (2) Random initialization of starting centers means that a global optimum may not be achieved in the clustering, and the resulting clustering depends on this initialization, varying with different starting centers. One possible advantage of the K-means algorithm, however, is that elements within a cluster typically remain more similar to one another using K-means than the "long" clusters of zero-order PH.

Table 8 Country and the corresponding year of the most recently-available data for each indicator, GDP per capita (GDP), Life Expectancy (LE), Infant Mortality (IM), GNI per capita (GNI)

Country	GDP	LE	IM	GNI	Country	GDP	LE	IM	GNI	Country	GDP	LE	IM	GNI
Afghanistan	2015	2016	2015	2010	Georgia	2015	2016	2015	2011	Palestine	2015	2016	2015	2005
Albania	2015	2016	2015	2011	Germany	2015	2016	2015	2011	Panama	2015	2016	2015	2011
Algeria	2015	2016	2015	2011	Ghana	2015	2016	2015	2011	Papua New Guinea	2015	2016	2015	2011
Angola	2015	2016	2015	2011	Greece	2015	2016	2015	2011					
Antigua and Barbuda	2015	2016	2015	2011	Grenada	2015	2016	2015	2011	Paraguay	2015	2016	2015	2011
					Guatemala	2015	2016	2015	2011	Peru	2015	2016	2015	2011
Argentina	2015	2016	2015	2011	Guinea	2015	2016	2015	2011	Philippines	2015	2016	2015	2011
Armenia	2015	2016	2015	2011	Guinea-Bissau	2015	2016	2015	2011	Poland	2015	2016	2015	2011
Australia	2015	2016	2015	2010						Portugal	2015	2016	2015	2011
Austria	2015	2016	2015	2011	Guyana	2015	2016	2015	2010	Qatar	2015	2016	2015	2011
Azerbaijan	2015	2016	2015	2011	Haiti	2015	2016	2015	2011	Romania	2015	2016	2015	2011
Bahamas	2015	2016	2015	2010	Honduras	2015	2016	2015	2011	Russia	2015	2016	2015	2011
Bahrain	2015	2016	2015	2010	Hungary	2015	2016	2015	2011	Rwanda	2015	2016	2015	2011
Bangladesh	2015	2016	2015	2011	Iceland	2015	2016	2015	2011	Saint Lucia	2015	2016	2015	2011
Barbados	2015	2016	2015	2009	India	2015	2016	2015	2011	Saint Vincent and the Grenadines	2015	2016	2015	2011
Belarus	2015	2016	2015	2011	Indonesia	2015	2016	2015	2011					
Belgium	2015	2016	2015	2011	Iran	2015	2016	2015	2009					
Belize	2015	2016	2015	2011	Iraq	2015	2016	2015	2011	Samoa	2015	2016	2015	2011
Benin	2015	2016	2015	2011	Ireland	2015	2016	2015	2011	Sao Tome and Principe	2015	2016	2015	2011
Bhutan	2015	2016	2015	2011	Israel	2015	2016	2015	2011					
Bolivia	2015	2016	2015	2011	Italy	2015	2016	2015	2011	Saudi Arabia	2015	2016	2015	2011
Bosnia and Herzegovina	2015	2016	2015	2011	Jamaica	2015	2016	2015	2011	Senegal	2015	2016	2015	2011
					Japan	2015	2016	2015	2011	Serbia	2015	2016	2015	2011
Botswana	2015	2016	2015	2011	Jordan	2015	2016	2015	2011	Seychelles	2015	2016	2015	2011
Brazil	2015	2016	2015	2011	Kazakhstan	2015	2016	2015	2011	Sierra Leone	2015	2016	2015	2011

Brunei	2015	2016	2015	2009	Kenya	2015	2016	2015	2011	Singapore	2015	2016	2015	2011
Bulgaria	2015	2016	2015	2011	Kiribati	2015	2016	2015	2011	Slovak	2015	2016	2015	2011
Burkina Faso	2015	2016	2015	2011	Korea Rep.	2015	2016	2015	2011	Republic				
Burundi	2015	2016	2015	2011	Kuwait	2015	2016	2015	2010	Slovenia	2015	2016	2015	2011
Cambodia	2015	2016	2015	2011	Kyrgyzstan	2015	2016	2015	2011	Solomon	2015	2016	2015	2011
Cameroon	2015	2016	2015	2011	Laos	2015	2016	2015	2011	Islands				
Canada	2015	2016	2015	2011	Latvia	2015	2016	2015	2011	South Africa	2015	2016	2015	2011
Cape Verde	2015	2016	2015	2011	Lebanon	2015	2016	2015	2011	Spain	2015	2016	2015	2011
Central	2015	2016	2015	2011	Lesotho	2015	2016	2015	2011	Sri Lanka	2015	2016	2015	2011
African Rep.					Liberia	2015	2016	2015	2011	Sudan	2015	2016	2015	2010
Chad	2015	2016	2015	2011	Libya	2015	2016	2015	2009	Suriname	2015	2016	2015	2010
Chile	2015	2016	2015	2011	Lithuania	2015	2016	2015	2011	Swaziland	2015	2016	2015	2011
China	2015	2016	2015	2011	Luxembourg	2015	2016	2015	2011	Sweden	2015	2016	2015	2011
Colombia	2015	2016	2015	2011	Macedonia	2015	2016	2015	2011	Switzerland	2015	2016	2015	2011
Comoros	2015	2016	2015	2011	FYR					Syria	2015	2016	2015	2010
Congo Dem.	2015	2016	2015	2011	Madagascar	2015	2016	2015	2011	Tajikistan	2015	2016	2015	2011
Rep.					Malawi	2015	2016	2015	2011	Tanzania	2015	2016	2015	2011
Congo Rep.	2015	2016	2015	2011	Malaysia	2015	2016	2015	2011	Thailand	2015	2016	2015	2011
Costa Rica	2015	2016	2015	2011	Maldives	2015	2016	2015	2011	Timor-Leste	2015	2016	2015	2010
Cote d'Ivoire	2015	2016	2015	2011	Mali	2015	2016	2015	2011	Togo	2015	2016	2015	2011
Croatia	2015	2016	2015	2011	Malta	2015	2016	2015	2010	Tonga	2015	2016	2015	2011
Cyprus	2015	2016	2015	2010	Mauritania	2015	2016	2015	2011	Trinidad and	2015	2016	2015	2011
Czech Rep.	2015	2016	2015	2011	Mauritius	2015	2016	2015	2011	Tobago				
Denmark	2015	2016	2015	2011	Mexico	2015	2016	2015	2011	Tunisia	2015	2016	2015	2011
Djibouti	2015	2016	2015	2009	Micronesia	2015	2016	2015	2011	Turkey	2015	2016	2015	2011
Dominica	2015	2016	2015	2011	Fed. Sts.					Turkmenistan	2015	2016	2015	2011
Dominican	2015	2016	2015	2011	Moldova	2015	2016	2015	2011	Uganda	2015	2016	2015	2011

(continued)

Table 8 (continued)

Country	GDP	LE	IM	GNI	Country	GDP	LE	IM	GNI	Country	GDP	LE	IM	GNI
Rep.					Mongolia	2015	2016	2015	2011	Ukraine	2015	2016	2015	2011
Ecuador	2015	2016	2015	2011	Montenegro	2015	2016	2015	2011	United Arab	2015	2016	2015	2011
Egypt	2015	2016	2015	2011	Morocco	2015	2016	2015	2011	Emirates				
El Salvador	2015	2016	2015	2011	Mozambique	2015	2016	2015	2011	United	2015	2016	2015	2011
Equatorial	2015	2016	2015	2011	Namibia	2015	2016	2015	2011	Kingdom				
Guinea					Nepal	2015	2016	2015	2011	United States	2015	2016	2015	2011
Eritrea	2015	2016	2015	2011	Netherlands	2015	2016	2015	2011	Uruguay	2015	2016	2015	2011
Estonia	2015	2016	2015	2011	New Zealand	2015	2016	2015	2010	Uzbekistan	2015	2016	2015	2011
Ethiopia	2015	2016	2015	2011	Nicaragua	2015	2016	2015	2011	Vanuatu	2015	2016	2015	2011
Fiji	2015	2016	2015	2011	Niger	2015	2016	2015	2011	Venezuela	2015	2016	2015	2011
Finland	2015	2016	2015	2011	Nigeria	2015	2016	2015	2011	Vietnam	2015	2016	2015	2011
France	2015	2016	2015	2011	Norway	2015	2016	2015	2011	Yemen Rep.	2015	2016	2015	2011
Gabon	2015	2016	2015	2011	Oman	2015	2016	2015	2010	Zambia	2015	2016	2015	2011
Gambia	2015	2016	2015	2011	Pakistan	2015	2016	2015	2011					

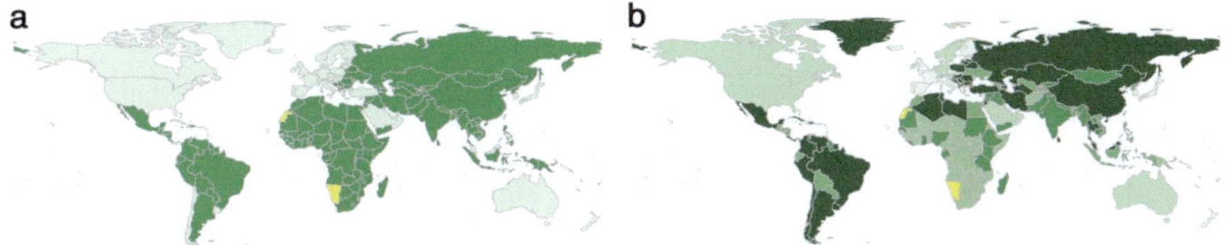

Fig. 7 World map depicting clusters found using K-means of $\mathbb{R}^2$ data: (**a**) $K = 2$ with cluster sizes 61 and 133; (**b**) $K = 6$ with cluster sizes 21, 25, 31, 32, 36, and 49. Shade corresponds to cluster size, where darker is larger. Yellow denotes countries missing from the data set because not all indicators are available

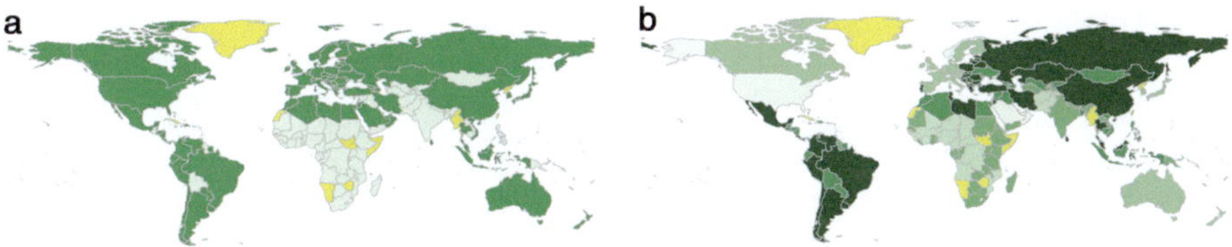

Fig. 8 World map depicting clusters found using K-means of $\mathbb{R}^4$ data: (**a**) $K = 2$ with cluster sizes 71 and 108; (**b**) $K = 6$ with cluster sizes 11, 21, 24, 37, 40, and 46. Shade corresponds to cluster size, where darker is larger. Yellow denotes countries missing from the data set because not all indicators are available

We perform the K-means algorithm using random initialization of cluster centers with two numbers of clusters $K = 2$ and 6 on the $\mathbb{R}^2$ and $\mathbb{R}^4$ data sets without geographic border information. The choice of $K = 2$ clusters is to consider whether the K-means algorithm separates countries into a first versus third world paradigm while the choice of $K = 6$ coincides with the six clusters used in the Gapminder representation (see Fig. 1) and in Sect. 4.1.

In Fig. 7a, we observe that the $K = 2$ clustering of the $\mathbb{R}^2$ data appears to follow what some may view as a first versus third world paradigm, grouping wealthier countries together. However, once more indicators have been added, this distinction starts to break down as the additional indicators reveal a more nuanced notion of similarity, see Fig. 8a. We observe that in Figs. 7b and 8b, when a more fine-grained approach is used to split the countries into more clusters, the clusters do not split along this paradigm or traditional continental divisions.

References

1. H. Adams, A. Tausz, Javaplex tutorial (2017), http://appliedtopology.github.io/javaplex/
2. G. Carlsson, Topology and data. Bull. Am. Math. Soc. **46**(2), 255–308 (2009)
3. M.K. Chung, P. Bubenik, P.T. Kim, Persistence diagrams of cortical surface data, in *Information Processing in Medical Imaging* (Springer, Berlin, 2009), pp. 386–397
4. Y. Dabaghian, F. Memoli, L. Frank, G. Carlsson, A topological paradigm for hippocampal spatial map formation using persistent homology. PLoS Comput. Biol. **8**(8), e1002581 (2012)

5. H. Edelsbrunner, J. Harer, Persistent homology – a survey. Contemp. Math. **453**, 257–282 (2008)
6. H. Edelsbrunner, J. Harer, *Computational Topology: An Introduction* (American Mathematical Society, Providence, 2010)
7. B.T. Fasy, F. Lecci, A. Rinaldo, L. Wasserman, S. Balakrishnan, A. Singh, Confidence sets for persistence diagrams. Ann. Stat. **42**(6), 2301–2339 (2014)
8. B.T. Fasy, J. Kim, F. Lecci, C. Maria, V. Rouvreau, The included GUDHI is authored by Clement Maria, Dionysus by Dmitriy Morozov, PHAT by Ulrich Bauer, Michael Kerber, and Jan Reininghaus. *TDA: Statistical Tools for Topological Data Analysis* (2017). R package version 1.5.1
9. R. Ghrist, Barcodes: the persistent topology of data. Bull. Am. Math. Soc. **45**(1), 61–75 (2008)
10. Global Burden of Disease Study 2013, *Global Burden of Disease Study 2013 (GBD 2013) Age-Sex Specific All-Cause and Cause-Specific Mortality 1990–2013* (Institute for Health Metrics and Evaluation (IHME), Seattle, 2014)
11. K. Heath, N. Gelfand, M. Ovsjanikov, M. Aanjaneya, L.J. Guibas, Image webs: computing and exploiting connectivity in image collections, in *Computer Vision and Pattern Recognition (CVPR), 2010 IEEE Conference on* (IEEE, New York, 2010), pp. 3432–3439
12. J. MacQueen et al., Some methods for classification and analysis of multivariate observations, in *Proceedings of the Fifth Berkeley Symposium on Mathematical Statistics and Probability, Oakland*, vol. 1 (1967), pp. 281–297
13. One World Nations Online, First, second and third world (2017), http://www.nationsonline.org/oneworld/third_world_countries.htm
14. J.A. Perea, J. Harer, Sliding windows and persistence: an application of topological methods to signal analysis. Found. Comput. Math. **15**(3), 799–838 (2015)
15. G. Singh, F. Memoli, T. Ishkhanov, G. Sapiro, G. Carlsson, D.L. Ringach, Topological analysis of population activity in visual cortex. J. Vis. **8**(8), 11 (2008)
16. B.J. Stolz, H.A. Harrington, M.A. Porter, The topological "shape" of brexit. CoRR, abs/1610.00752 (2016)
17. The Gapminder Foundation, Gapminder tools (2017), http://www.gapminder.com/tools. Accessed 29 Jan 2017
18. The geonames geographical database, http://www.geonames.org. Accessed 28 Jan 2017
19. The World Bank, World development indicators. Gni per capita, ppp (2012), http://data.worldbank.org/indicator/NY.GDP.PCAP.KD
20. The World Bank, World development indicators. Gdp per capita (2013), http://data.worldbank.org/indicator/NY.GDP.PCAP.KD
21. C.M. Topaz, L. Ziegelmeier, T. Halverson, Topological data analysis of biological aggregation models. PLoS ONE **10**(5), e0126383 (2015)
22. UNICEF, Child mortality estimates (2015), http://childmortality.org
23. A. Zomorodian, G. Carlsson, Computing persistent homology. Discret. Comput. Geom. **33**(2), 249–274 (2005)

Cluster Identification via Persistent Homology and Other Clustering Techniques, with Application to Liver Transplant Data

Berhanu A. Wubie, Axel Andres, Russell Greiner, Bret Hoehn, Aldo Montano-Loza, Norman Kneteman, and Giseon Heo

Abstract Clustering, an unsupervised learning method, can be very useful in detecting hidden patterns in complex and/or high-dimensional data. Persistent homology, a recently developed branch of computational topology, studies the evolution of topological features under a varying filtration parameter. At a fixed filtration parameter value, one can find different topological features in a dataset, such as connected components (zero-dimensional topological features), loops (one-dimensional topological features), and more generally, k-dimensional holes (k-dimensional topological features). In the classical sense, clusters correspond to zero-dimensional topological features. We explore whether higher dimensional

B. A. Wubie
Department of Mathematical and Statistical Sciences, University of Alberta, Edmonton, Canada
e-mail: wubie@ualberta.ca

A. Andres
Service de chirurgie viscérale et transplantation, Hôpitaux Universitaires de Genève, Geneva, Switzerland
e-mail: Axel.Andres@hcuge.ch

R. Greiner
Computing Science, University of Alberta, Edmonton, Canada
e-mail: rgreiner@ualberta.ca

B. Hoehn
Alberta Innovates Centre for Machine Learning, Edmonton, Canada
e-mail: bhoehn@ualberta.ca

A. Montano-Loza
Hepatology, Department of Medicine, University of Alberta Hospital, Edmonton, Canada
e-mail: montanol@ualberta.ca

N. M. Kneteman
Transplantation surgery, Dept. of Surgery, University of Alberta Hospital, Edmonton, Canada
e-mail: kneteman@ualberta.ca

G. Heo (✉)
School of Dentistry, University of Alberta, Edmonton, Alberta, Canada
e-mail: gheo@ualberta.ca

© The Author(s) and the Association for Women in Mathematics 2018
E. W. Chambers et al. (eds.), *Research in Computational Topology*, Association for Women in Mathematics Series 13, https://doi.org/10.1007/978-3-319-89593-2_9

homology can contribute to detecting hidden patterns in data. We observe that some loops formed in survival data seem to be able to detect outliers that other clustering techniques do not detect. We analyze patterns of patients in terms of their covariates and survival time, and determine the most important predictor variables in predicting survival times of liver transplant patients by applying a random survival forest.

1   Introduction

Clustering is the partitioning of objects in a set into distinct clusters with specific traits that distinguish them from other clusters. Clustering is a critically important and well-studied element of statistics and is effective in revealing unknown patterns in data or geometric objects. Unlike classification, the number of clusters in clustering analysis is unknown: As such, this method falls under the category of unsupervised learning. There are several methods in clustering, including k-means, hierarchical, spectral, geometric graph, and density-based methods. For a general overview and example of clustering, see [1, 24]. For most clustering methods, one should have some prior knowledge regarding the number of clusters or, alternatively, some method of estimating the number of clusters present.

Persistent homology (PH), a branch of computational topology, combines the differentiating power of geometry and the classification power of topology [6]. Persistent homology has recently been developed [10, 31] and popularized as part of topological data analysis [5, 12]. Persistent homology is used to detect true signal in geometrical objects by varying a parameter rather than finding an optimal parameter value that yields the most accurate representation of the signal. In various datasets, this has shown to be effective in discerning true signal from noise [4, 30].

One of the main objectives of this article is to assess whether persistent homology can identify certain patterns in survival data sets. With a simple motivating example, Fig. 1 illustrates clustering results based on persistent homology (explained in Sect. 2), partitioning around medoids, and spectral clustering (explained in Sect. 5). Ben-Hur et al. [2] applied a support vector clustering technique to concentric rings and were able to detect three rings as clusters (see Fig. 3 in their article): We replicate this data set with two concentric rings in Fig. 1. Analysis using persistent homology shows two persistent loops and this corresponds to the two clusters in [2]. The clusters obtained from partitioning around medoids, spectral clustering, and persistent homology are also compared in Fig. 1. It is interesting to note that the clustering pattern based on partitioning around medoids is similar to that based on PH at a certain parameter value, a topic we expand upon in Sect. 2.

This study uses data from the Scientific Registry of Transplant Recipients (SRTR). The SRTR data system includes information on all donors, waitlisted candidates, and transplant recipients in the United States, submitted by members of the Organ Procurement and Transplantation Network (OPTN). The Health Resources and Services Administration (HRSA), US Department of Health and Human Services, provides oversight to the activities of OPTN and SRTR contractors.

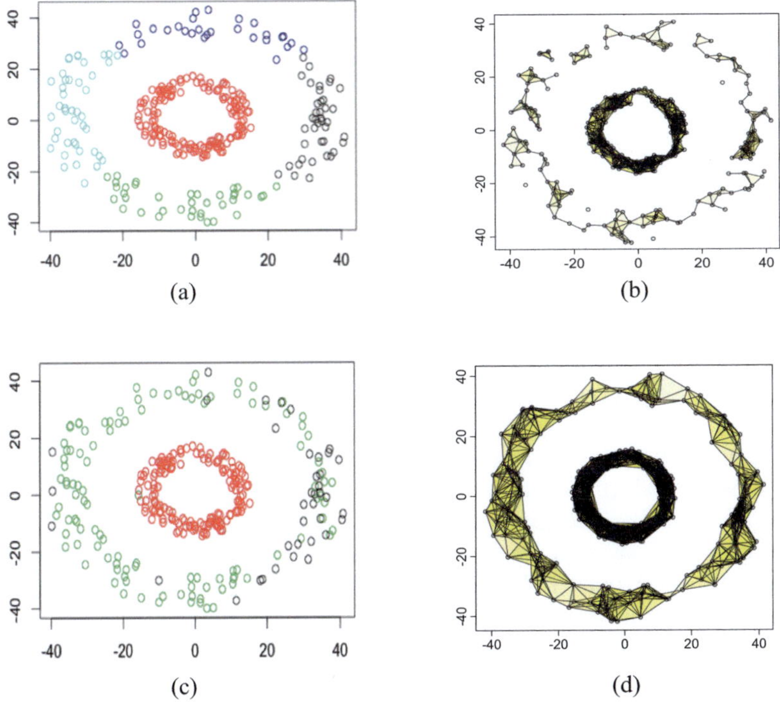

Fig. 1 Simulated dataset to replicate Ben-Hur et al.'s concentric ring dataset [2]. (**a**) Partitioning around medoids shows five clusters. (**b**) Persistent homology simplicial complex present at scale parameter $\epsilon = 5.5$ indicates seven clusters. (**c**) Spectral clustering indicates two clusters in red and green, while points in black do not belong to these two clusters. (**d**) Persistent homology simplicial complex for scale parameter $\epsilon = 10$ indicates two persistent clusters

The SRTR data set consists of a total of 8361 alcoholic patients who received liver transplants and are under follow-up to be assessed and classified into groups based on their quality of life posttransplant. In classifying patients, we consider different clinical and demographic factors associated with failure of liver transplants in alcoholic patients. Although there are many clinical, demographic, and socioeconomic factors associated to transplant failure (such as death of a patient), we considered those factors associated with the time to patient death after transplant. Demographic factors considered are the recipient's gender, age, blood types, height, and weight, while the clinical factors are creatinine, bilirubin, and albumin levels, and the transplant factor considered is the cold ischemia time of the organ. We present results based on the analysis of patients' data. We will use failure time and survival time interchangeably for the time (in days) from liver transplant until death. Recall that censoring time indicates incomplete information regarding a patient's event time, occurring when a participant drops out of the study or when participant survival time is greater than the time of the last follow-up. In our case, the response variable is the survival and censoring times of alcoholic patients after receiving liver transplants. Methods in survival analysis are typically used in the analysis of this type of censored data.

The most common methods include the classical log-rank test and the Cox proportional hazard model, and also recent approaches such as random survival forest (RSF), a non-parametric method for ensemble estimation constructed from classification trees for survival data and is used as an alternative method for better survival prediction and ranking the importance of covariates associated with it [3]. We first apply RSF to determine the most influential covariates in determining survival time, and we then explain the survival time clustering patterns in terms of these most important covariates. Due to missing information, however, we only use data for 7154 patients in RSF. Furthermore, due to computational issues in implementing persistent homology using the R-package TDA (topological data analysis) [11] for large datasets, for this study a personal laptop is able handle persistent homology calculations up to $n = 1000$. Thus, our analysis by RSF is based on the total sample, whereas the results of cluster analysis via persistent homology are based on samples of 1000 cases. These samples are randomly selected proportional to the number of patients failed (67%) and censored (33%).

We close introduction with a description of sections to follow. In Sect. 2, we briefly introduce the background of persistent homology. In Sect. 3, we briefly introduce random survival forest and determine the variables most important in predicting survival time. Following this, we provide a brief explanation of the dissimilarity measure used in this study in Sect. 4 and two clustering methods in statistics in Sect. 5. Section 6 describes the distribution of survival times, censoring, and covariates. In Sect. 7, we compare the clusters obtained by these three different methods. We conclude our project with recommendations for future study in Sect. 8. Clustering methods, RSF, and persistent homology are not new, and thus are well explained in many textbooks [3, 9, 17, 24] for example, so we do not explain them in detail in this manuscript. However, for the sake of reader background, we briefly explain RSF and its algorithm in an appendix to this paper.

2 Brief Review of Persistent Homology

Let $S = \{X_1, \ldots, X_n\}$ be sample points taken randomly from a manifold X. We want to obtain the homology of X using S. First, we cover each point by an ϵ-ball $B_\epsilon(X_i)$, from which we can construct simplicial complexes. Well-known simplicial complex constructions are the Čech and Vietoris–Rips [7, 9] varieties. In this article we apply the Vietoris–Rips construction.

One could try to find a value of ϵ such that X is homotopy equivalent to $\bigcup_i B_\epsilon(X_i)$ so that both have the same homology. Rather than searching for an appropriate ϵ to capture the "true" topology of a space, however, computational topologists developed persistent homology to study the family of simplicial complexes generated as the parameter ϵ varies [10, 15, 31].

We now provide a brief mathematical definition of persistent homology. Let $(K_i)_1^m$ be, a nested sequence filtration, of complexes of S for an increasing sequence of parameter values $(\epsilon_i)_1^m$. In this setting, there is a natural inclusion map

$$\emptyset = K_0 \overset{\iota}{\hookrightarrow} K_1 \overset{\iota}{\hookrightarrow} K_2 \overset{\iota}{\hookrightarrow} \cdots \overset{\iota}{\hookrightarrow} K_m. \tag{1}$$

The kth homology group of the iterated inclusions $\iota_k : H_k(K_i) \rightarrow H_k(K_j)$ for all $i < j$ reveals which kth order topological features persist. The persistence complex $\mathscr{C} = \{C_*, f^i\}$ is a family of chain complexes $\{C_*^i\}$ with chain maps $f^i : C_*^i \rightarrow C_*^{i+1}$. The (i, j) persistent homology of $\mathscr{C}$ is defined to be the image of the induced homomorphism $f_*^{i,j} : H_*(C_*^i) \rightarrow H_*(C_*^{i+j})$. Within a filtration, persistent homology captures all topological characteristics in a complete, discrete invariant. This invariant is expressed as a multiset of intervals, that is, a barcode.

Figure 2 shows a simple illustration of persistent homology. We consider 300 randomly selected points from two 'crossed' ellipses and build Rips complexes at various ϵ values. As the radius increases, more disks overlap and the simplicial complex evolves. Several loops are formed (born) and are quickly filled in (so as to become a disk which is homotopy equivalent to a point and thus not interesting from a topological standpoint), and these are considered to be noise. The birth and death times of each topological feature can be denoted by b and d, so that the duration of that feature's survival is represented as a half interval $[b, d)$. The multiset of intervals of the form $[b, d)$ is called a barcode [6, 18]. Long intervals represent topological features that persist over a wide range of scales and thus likely indicate "true" signal in the data. On the other hand, short intervals in the barcode represent features that are only present for a small range of scales and likely exist only due to random noise in the data. In topology, a zero-dimensional feature can be viewed as clusters formed by points, which corresponds to clusters in classical statistics known as single linkage clustering. The loop is considered a one-dimensional feature in topology. In general, a k-dimensional hole is called a k-dimensional feature. The number of k-dimensional holes is denoted as the kth Betti number, β_k.

We intuitively illustrate the basic concept of persistent homology with an example in Fig. 2. In Fig. 2, we observe that the number of clusters decreases as ϵ increases. For example, four clusters are present at $\epsilon = 0.28$, and two clusters at $\epsilon = 0.38$. We also see loops formed by chains of edges (1-simplicies) and triangles (2-simplicies)); for instance, three small loops are formed at $\epsilon = 0.28$. These three loops disappear while three new larger loops and one new small loop appear when ϵ reaches 0.38.

We investigate the observations that generate the five loops (1-cycles) that persist over the widest range of scales, that is, correspond to the longest barcode intervals of (d) in Fig. 2. The generators for these cycles are not unique and those presented in (f) are one of many possible generators for each loop. The points forming representative cycles for these loops are shown in five bright colors, while those points shown in black do not belong to any of the chosen five representative cycles. The points in (c) and (f) in Fig. 2 are vertices in the boundaries of loops found by Dionysus [25]. Dionysus finds these cycles through a matrix reduction algorithm.

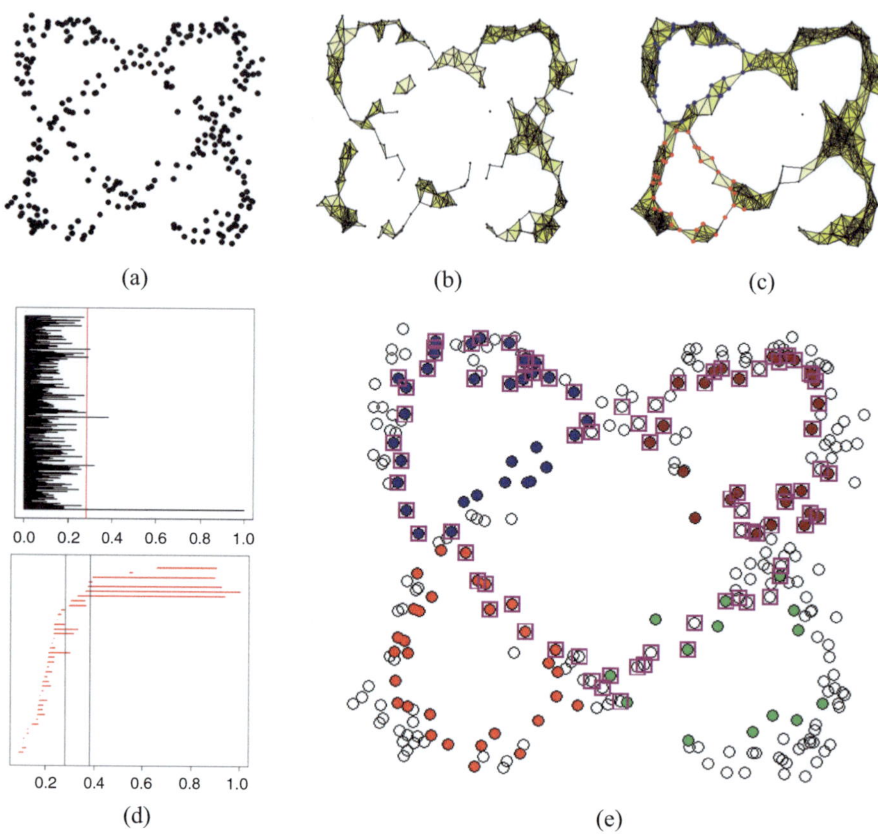

Fig. 2 (**a**) 300 points randomly sampled from two overlapping ellipses; at scale parameter $\epsilon = 0.0$, there are $\beta_0 = 300$ connected components. (**b**) At $\epsilon = 0.28$, points are connected by edges and triangles, resulting in four connected components ($\beta_0 = 4$) and three loops ($\beta_1 = 3$). (**c**) Rips complex at $\epsilon = 0.38$. Generators for two representative loops are shown in red and blue. (**d**) Barcode for dimension 0 and 1 in black and red, respectively. The vertical line at $\epsilon = 0.28$ in the β_0-barcode indicates four connected components. At $\epsilon = 0.28$, three loops exist and at $\epsilon = 0.38$, four loops. (**e**) Five representative loops are indicated using five different colors (red, blue, green, brown, and magenta) corresponding to top five longest intervals (bars) in (**d**). The generators for the four loops in the corners do not overlap, but the generators for the largest loop indicated by magenta squares show that some of the vertices belong to other loops as well. Points shown in black are not part of generators for any of five loops. (**a**) Point cloud at $\epsilon = 0.0$. (**b**) Features at $\epsilon = 0.28$. (**c**) Generators for two loops. (**d**) β_0 & β_1-barcode. (**e**) Five representative loops

3 Determining the Most Important Covariates Using Random Survival Forest

Random survival forest is an ensemble method that combines a number of trees by taking the same number of bootstrap samples from the original data and growing a tree on each. It is a method designed for survival data. The individual trees in a random survival forest are constructed using a randomly selected subset of predictors for splitting the root nodes into new daughter nodes at each split. A root node is split into daughter nodes using a splitting criteria on that predictor that maximizes survival differences across daughter nodes [19, 22]. We grow the trees to full size on each bootstrap sample under the constraint that a terminal node should have no less than a pre-specified node size of unique events (i.e., deaths). The individual trees in a random survival forest are not pruned and used for decision; instead, decisions are based on all trees grown in the bootstrap samples by a similar procedure, from which we can generate a forest. First, for each tree grown from a bootstrap data set we estimate the cumulative hazard function for the tree. This is accomplished by grouping hazard estimates for each terminal node.

As shown in Fig. 3, to compute the cumulative hazard estimate for an individual l not part of the bootstrap sample (i.e., out of bag data) with predictor X_l, simply drop X_l down the trees in which individual l is out of bag (OOB). The terminal node for l yields the desired estimator for each tree determined in the first step [19, 23]. Consequently, the OOB ensemble cumulative hazard estimator is obtained by averaging the cumulative hazard estimate over those bootstrap trees in which individual l is excluded (i.e., those datasets in which l is OOB).

We first apply RSF for the entire data set, 7154 alcoholic patient and grow trees based on the nine patient covariates. To run the RSF algorithm we grow 1000 random trees of large size and three covariates taken randomly for splitting a parent node into different daughter nodes of similar survival experience until the minimum terminal node size three patients is reached, a point in which we would stop further splitting a node. The prediction error in using RSF for the overall grown trees is found to be 33.26%. The prominent advantage of using RSF over the standard cox model is that we can identify and rank the most important covariates used for growing trees in a forest so as to have reliable predictive ability of its survival for the OOB patients who receive liver transplant or new patients waiting to undergo liver transplant. As a result, effectiveness of covariates in predicting survival times of liver transplant patients was ranked from most to least; creatinine, cold ischemia time of the organ, bilirubin, albumin, height, age, weight, blood type, and gender (see Table 19 in Appendix).

4   Dissimilarity Measure for Cluster Analysis

In data clustering using persistent homology, the input is a proximity index or dissimilarity measure. Therefore, before applying persistence-based clustering, we must obtain a distance measure for the dataset. A procedure for distance measure computation for cluster analysis is described by Kaufman and Rousseeuw [24]: A brief summary for their method is given below.

First, the dataset used for clustering is arranged into the following structures of n items with p attributes, that is the dimension of the data should be an $n \times p$ objects-by-attributes matrix, where rows represent items or objects, and columns

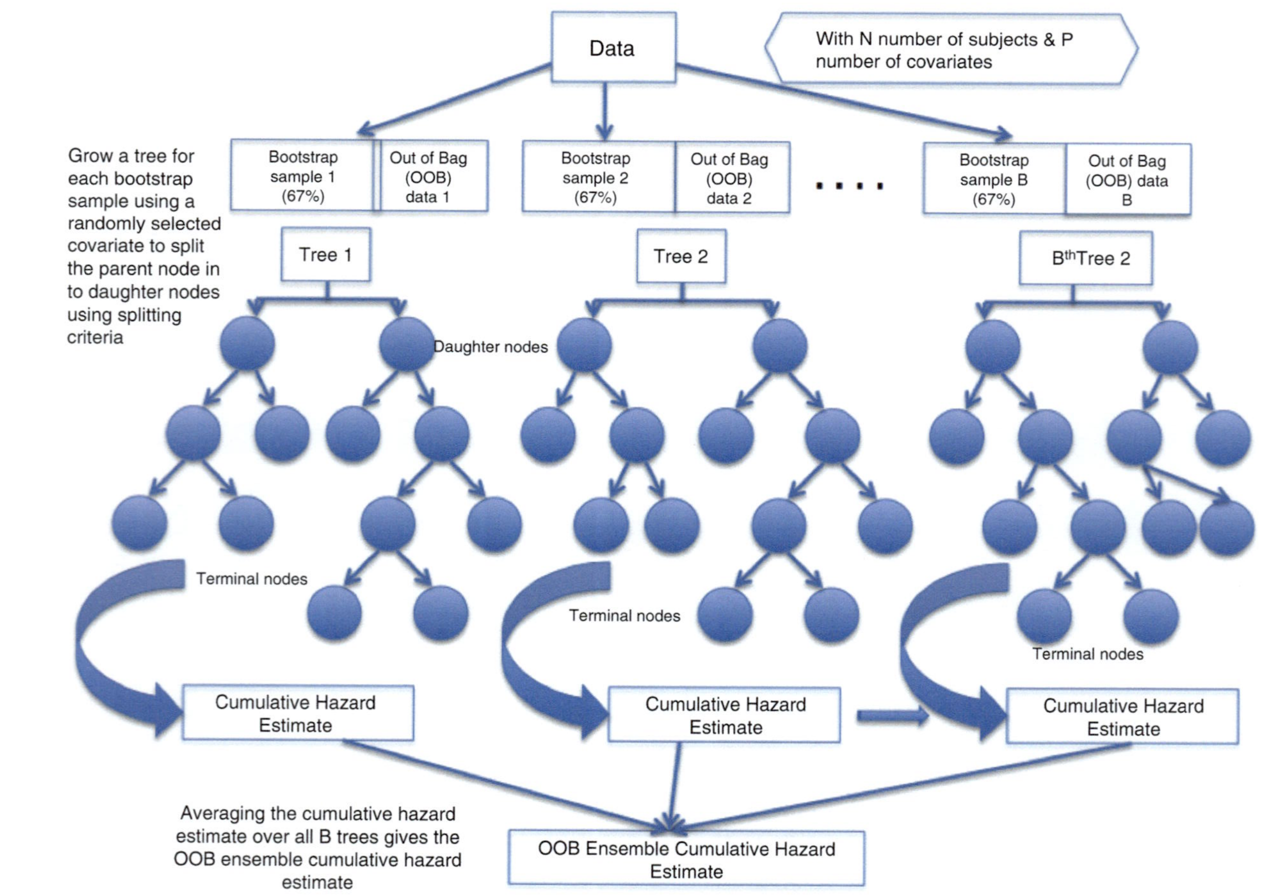

Fig. 3 A schematic illustration of the random survival forest (RSF) algorithm

represent the variables associated with each object. If the data is arranged in this way we can compute the $n \times n$ dissimilarity matrix, using the DAISY auxiliary program in the cluster package for R. This program computes the dissimilarity matrix for items using their given attributes. The dissimilarity matrix computed from the dataset, written as $d\left(X_i, X_j\right) = d\left(X_j, X_i\right)$, measures the difference or dissimilarity between the objects X_i and X_j, both in $\mathfrak{R}^p$. The advantage of using the DAISY program is that it can handle numerous variable types, including nominal, ordinal, asymmetric binary, and ratio-scaled, in the computation of the dissimilarity measure between objects. In calculating the distance measure, the DAISY program has three options—the Euclidean (default) and Manhattan, both for dataset with all metric attributes; and the third, an extension of the above two developed by Gower (1971) that incorporates and handles attributes of mixed types in a dataset for the calculation of a proximity measure. The resulting dissimilarity measure $d\left(X_i, X_j\right) = d\left(i, j\right)$ developed by Gower can be defined as [24]

$$d\left(i, j\right) = \frac{\sum_{f=1}^{p} \delta_{ij}^{(f)} d_{ij}^{(f)}}{\sum_{f=1}^{p} \delta_{ij}^{(f)}},$$

where $d_{ij}^{(f)}$ is the contribution of variable f to $d\left(i, j\right)$, which depends on variable f's type:

- If f is binary or nominal: $d_{ij}^{(f)} = 0$ if $X_{if} = X_{jf}$, and $d_{ij}^{(f)} = 1$ otherwise;
- If f is interval-scaled: $d_{ij}^{(f)} = \frac{|X_{if} - X_{jf}|}{max_h X_{hf} - min_h X_{hf}}$;
- If f is ordinal or ratio-scaled: first, rank the ith object with respect to variable f, denote the rank as r_{if}, then compute $z_{if} = \frac{r_{if} - 1}{max_h r_{hf} - 1}$, and treat these z_{if} as interval-scaled variable and compute $d_{ij}^{(f)}$ as above.

Furthermore, the weight $\delta_{ij}^{(f)}$ of the variable f is also affected by f's type as a variable:

- $\delta_{ij}^{(f)} = 0$ if X_{if} or X_{jf} is missing;
- $\delta_{ij}^{(f)} = 0$ if $X_{if} = X_{jf} = 0$ and variable f is asymmetric binary; and
- $\delta_{ij}^{(f)} = 1$, otherwise.

The output from DAISY is an object of the class dissimilarity, and can be used as input for several other clustering functions, like partitioning around the medoid, spectral clustering, and clustering using persistent homology. Hence, the same dissimilarity measure computed from DAISY was used as an input for the clustering techniques applied in this study.

5   Brief Description of Clustering Methods

A general idea of clustering is to find k representatives in the set and then divide the remaining points into classes based on which representative each is closest to. The representative points are called cluster centers, centroids, or prototypes. Consider n observations $\mathbf{x} = \{x_1, \ldots, x_n\}$ taken from $\Re^p$. Clustering in k-means aims to find optimal k partitions, $\mathbb{C} = \{C_1, \ldots, C_k\}$ of the n observations by minimizing the sum of squared distances between each point and the center of its respective cluster, namely,

$$\operatorname*{argmin}_{\mathbb{C}} \sum_i^k \sum_{x_i \in C_i} d(x_i, \eta_i), \tag{2}$$

where d is L_2 norm, and η_i is centroid of the ith cluster. It is well known that k-means clustering is sensitive to outliers. Because of this sensitivity to outliers, researchers have proposed k-medians clustering, where the centroid is the median instead of the mean, obtained by applying the L_1 norm in Eq. (2).

We observe that means and medians do not necessarily belong to the original set $\mathbf{x}$ of the observations. For this reason, k-medoids clustering was proposed, where a medoid is defined as the point of the cluster for which the average dissimilarity to all the points of the cluster is minimal [24]. Partitioning around medoids (PAM) is among many k-medoids clustering algorithms, and one that divides the dataset into k clusters, where the integer k needs to be specified by the user. Typically, the user runs the algorithm for a range of k-values. For each k, the algorithm carries out the clustering and also yields a "quality index", which allows the user to select a value of k afterward. PAM in [24] is considered to be most powerful [26, 28]. The distance in (2) can be any dissimilarity measure and is not restricted to the L_1 or L_2 norm.

For each element x_i, we calculate the average dissimilarity of x_i within its cluster, denoted w_i, and the average dissimilarity of x_i in each cluster in which x_i is not a member, denoted $b_{i\ell}$. Let $b_i = \min_\ell b_{i\ell}$. The silhouette for x_i is defined as

$$S_i(\mathbb{C}) = \frac{b_i - w_i}{\max(b_i, w_i)}. \tag{3}$$

The average silhouette is known to be a good measure of the strength of clustering results. That is, it determines how well each object lies within its cluster. A high average silhouette width indicates a good clustering. [14, 28].

The ideas underlying spectral clustering are related to eigenvalues and eigenvectors. We construct an undirected graph G with n vertices that corresponds to the n observations $x_1, \ldots, x_n$. Two vertices x_i and x_j are connected if the weight W_{ij} between these two vertices is larger than zero. The simplest weights that can be used in a weight matrix W is $\{0, 1\}$, that is, $W_{ij} = \mathbf{1}(||x_i - x_j|| < \epsilon)$, for some value ϵ, where $\mathbf{1}$ denotes an indicator function. Another general example of a weight matrix is $W_{ij} = e^{-||x_i - x_j||^2/\sigma^2}$. The graph Laplacian is defined as $L = D - W$, where D

is the n by n diagonal matrix with entry $D_{ii} = \sum_j W_{ij}$. The graph Laplacian L has interesting properties, namely, (1) L is symmetric and positive semi-definite, (2) the smallest eigenvalue of L is 0 and its corresponding eigenvector is a unit vector, (3) L has n nonnegative, real-valued eigenvalues $0 = \lambda_1 \leq \ldots \leq \lambda_k$, and (4) the number of connected components in the graph G is the number of zero eigenvalues. The corresponding eigenvectors are orthogonal, and each eigenvector is constant over one of the connected components of the graph. One advantage of spectral clustering is that we need not specify the number of clusters. Some researchers define the graph Laplacian slightly differently, as the symmetrized Laplacian $D^{-1/2} W D^{-1/2}$ or the random walk Laplacian, $D^{-1} W$ or $I - D^{-1} W$. For a good introduction to spectral clustering, see [29]. In our illustrative example above and for our data analysis in Sect. 6.2, Fig. 5a, we applied the eigengap heuristic, a particular method designed for spectral clustering techniques that can be applied to graph Laplacians. The goal of this technique is to choose the number of clusters k such that all eigenvalues $\lambda_1 \leq \ldots \leq \lambda_k$ are very small and λ_{k+1} relatively large.

The clusters formed by PAM, spectral, and persistent homology (PH) for the synthetic data are presented in Fig. 1. From the clusters presented, we can see that, for the two bands, PAM forms five clusters, spectral clustering forms two clusters while leaving several points (shown in black) unassigned to any cluster, and persistent homology forms two clusters at scale parameter value $\epsilon = 10$. PAM clustering considers each data point as a centroid in forming clusters and applies the dissimilarity matrix of the given data set to minimize the sum of distances. The drawback of PAM relative to other methods is that data points are clustered into many clusters as a result of using the medoids (median) as a centre for clustering. Spectral clustering, on the other hand, clusters data points by finding an optimal partition based on an affinity graph from the data set and clusters the data points into two clusters with some noise points.

For a closer look at representative loops (1-cycles) in the synthetic data, see Fig. 2. In the sections below, we compare patient clustering patterns obtained from clustering methods with the representative cycles found by persistent homology. We identify clusters found using PAM, spectral clustering, and PH0 (persistent homology dimension 0) and also representative loops found by PH1 (persistent homology dimension 1), which we refer to as representative loops or, simply, loops. In the following section, we discuss distribution of alcoholic patients in each cluster formed by PAM, Spectral, PH0, and PH1, median survival times, and corresponding median values of the three most important covariates chosen by RSF, namely, creatinine level, cooling time of the organ, and bilirubin level.

6 Discussion of Clusters and Representative Loops Obtained from PAM, Spectral, PH0, and PH1

6.1 Clustering by PAM

In PAM, the maximum number of clusters is determined by considering the value that maximizes the average Silhouette width (Fig. 4). PAM clustering, as shown above in Fig. 4b, gave seven different clusters with different median survival times (Table 1). From the table, we can see that cluster two is characterized as having the highest median survival time after liver transplant and patients in this cluster showed a better survival experience after transplant than patients in cluster three (Log-Rank p-value = 0.0494). The next best cluster in terms of survival rate is cluster four, as patients in cluster four had a median survival time of 2352 days after receiving a liver transplant. The cluster that showed the shortest median survival time and poorest survival experience is cluster five. That is about 66.3% of the patients in cluster five had a median survival time of 1473 days after receiving transplant. In other words, patients in cluster five had a poor survival experience relative to patients in other clusters, as shown in Fig. 4c. PAM clustering also showed that, of those patients considered in the study, more than 50% were found in the first three clusters, with 28.6% in the first cluster, about 10.1% in the second, and about 34.7% of them in the third. In all these clusters, more than 50% of those patients in each cluster died after receiving the transplant at different times and no significant differences in their survival experiences exist.

The clusters formed by PAM showed different patterns of covariates: the lowest median bilirubin value is registered in cluster seven, 3.20 (95% CI, 0.69, 22.68), and the highest in cluster five, that is, 6.85 (95% CI, 0.73, 34.14). Considering the creatinine level of patients, the minimum median is registered in cluster four at 1.00 (95% CI, 0.50, 4.00). In a similar trend as we can see from Table 2 that the lowest of the median cold ischemia time for the organ (Cold_Isch) is found in those patients that are in cluster six, 6.00 (95% CI, 2.43, 17.45).

The highest median survival time is observed in cluster two, and the lowest median survival time in cluster five. Each of the three covariates is distributed similarly in their respective clusters. For instance, patients in cluster five are found to have the highest median bilirubin level and the second lowest median creatinine level among all seven clusters (see Table 2).

6.2 Spectral Clustering

Spectral clustering was also used in clustering alcoholic patients that received liver transplants. The eigenvalue plot displayed the optimum number of clusters to be considered: As seen from Fig. 5a, this optimum number is five. Figure 5a shows the visualization of the five clusters in a 2D multidimensional scaling plot,

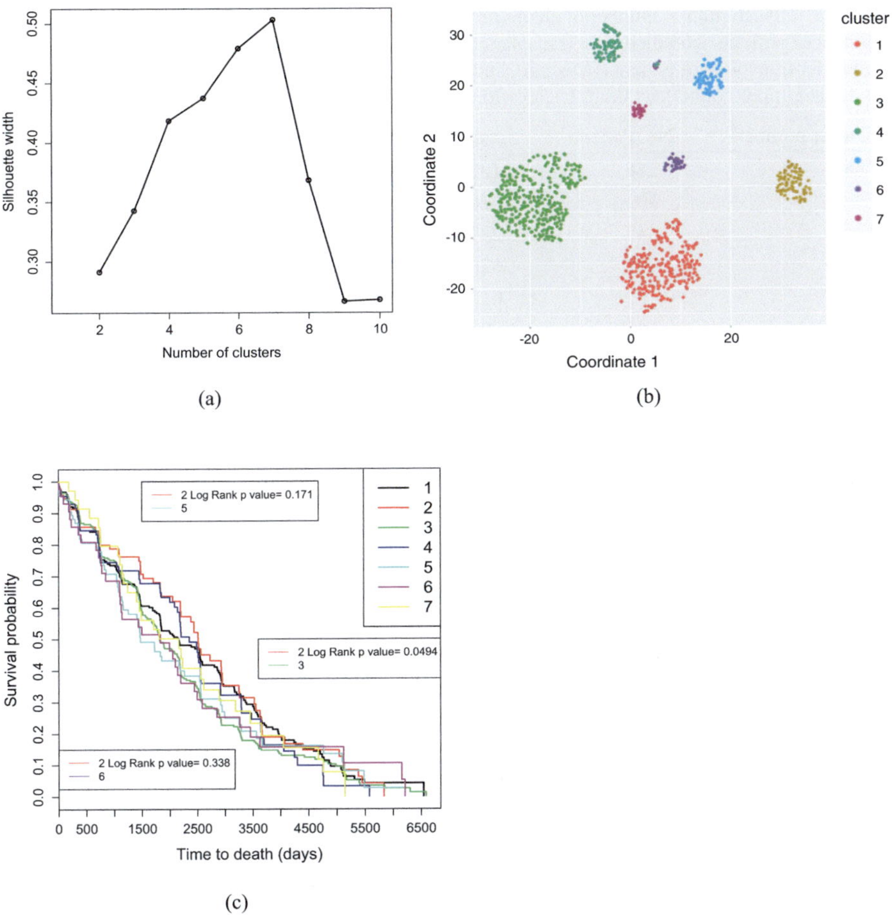

Fig. 4 (**a**) Average silhouette width plot with number of clusters to determine the maximum number of clusters in PAM, cluster with high average silhouette width ($k = 7$). (**b**) 2D multidimensional scaling scatter plot representation of the seven clusters of alcoholic patients formed by PAM. (**c**) Survival curves for the corresponding seven clusters formed by PAM and log-rank p-value comparison of cluster survival curves for patients in clusters 2 and 3 (p-value = 0.171), clusters 2 and 5 (p-value = 0.0494), and clusters 2 and 6 (p-value = 0.338). (**a**) Silhouette width plot. (**b**) Cluster scatter plot. (**c**) Survival curves

with 292 patients in cluster one, of which 72.9% patients died after receiving the transplant. Similarly, there are 121, 300, 201, and 86 patients in clusters two, three, four, and five, respectively, and more than 60% of the patients in all clusters died after receiving transplants. The clusters possess different median survival times, the maximum being 2197 days for those patients in cluster four, and the minimum being 1722 days after transplant for those patients in cluster five, as shown in Table 3. Spectral clustering as seen in Fig. 5b shows significant differences in the survival experience of patients in clusters four and five (Log-Rank p-value = 0.0273). Those

Table 1 Distribution of number of alcoholic patients who received liver transplant (No. patients), number of patients who died after transplant (No. events), and the corresponding median survival time after transplant of patients (Median), and its 95% confidence interval (lower confidence limit, LCL, and upper confidence limit, UCL) with respect to PAM clusters

Cluster number	No. patients	No. events	Median	LCL	UCL
One	286	182(63.9)	2160	1832	2587
Two	101	62(61.4)	2508	2192	2949
Three	347	239(68.9)	1815	1674	2172
Four	91	62(68.1)	2352	2139	2560
Five	92	61(66.3)	1473	1163	2540
Six	46	36(78.2)	1827	1108	2490
Seven	37	30(81.1)	2161	1403	3179
Total	1000	672			

Table 2 Median values of the most important covariates: bilirubin level (Bilirubin), creatinine level (Creatinine), and cold ischemia time for the organ (Cold_Isch) of alcoholic patients after receiving liver transplant, and the corresponding 95% confidence interval (lower confidence limit, LCL, and upper confidence limit, UCL) within each cluster formed by PAM

Cluster number	Bilirubin	LCL	UCL	Creatinine	LCL	UCL	Cold_Isch	LCL	UCL
One	3.75	0.80	32.24	1.30	0.60	4.26	7.00	3.00	15.88
Two	4.40	0.95	31.70	1.20	0.60	4.25	6.80	3.10	13.20
Three	5.00	0.80	43.37	1.27	0.50	4.70	7.00	2.19	13.64
Four	5.60	0.90	31.85	1.00	0.50	4.00	7.00	1.00	13.71
Five	6.85	0.73	34.14	1.10	0.43	4.00	6.95	2.03	13.00
Six	3.35	0.41	49.96	1.30	0.70	4.00	6.00	2.43	17.45
Seven	3.20	0.69	22.68	1.19	0.59	4.13	6.23	2.85	13.05

patients in cluster four have a better survival experience than patients in cluster five. In other words, patients in cluster five showed poor survival experience compared to patients in the remaining clusters.

Spectral clustering has already identified five different clusters of which the fifth cluster is known to include those patients that have extremely high median bilirubin levels, namely 31.75 (95% CI, 20.91, 54.75) on average. Aside from having the highest median bilirubin level, patients in this cluster also display the highest median creatinine level of 3.00 (95% CI, 0.52, 5.71) but showed the minimum median cold ischemia time of the organ (cold_isch) compared to the other clusters, as shown in Table 4. On the other hand, patients in cluster two are known to have the smallest median bilirubin level, 3.45 (95% CI, 0.83, 16.95), and the second smallest median creatinine level, 1.20 (95% CI, 0.59, 4.00), but the second highest median cold ischemia time, 7.00 (95% CI, 2.23, 13.40), relative to the remaining clusters.

The median survival times in the obtained spectral clusters are close to each other with the exception of cluster 5, in which the median survival time is lowest and significantly different from that of other clusters (Table 3). The values of the

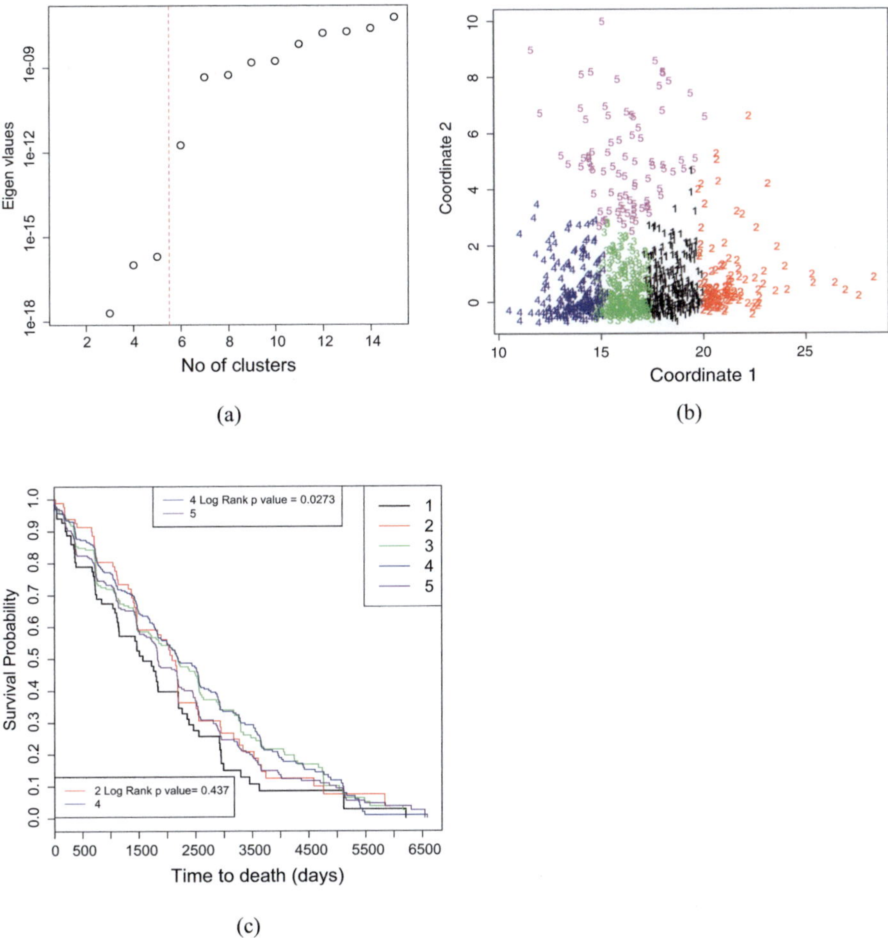

Fig. 5 (**a**) A plot of eigen values with the number of clusters to choose the optimal number of clusters in spectral clustering, most stable clustering is usually given by the value that maximizes the eigen-gap (difference between consecutive eigenvalues, $k = 5$). (**b**) 2D multidimensional scaling scatter plot representation of the five clusters formed by spectral clustering for alcoholic patients receiving liver transplant. (**c**) Survival curves for the corresponding five clusters formed by spectral clustering and log-rank p-value comparison of cluster survival curves for alcoholic patients in clusters 2 and 4 (p-value $= 0.437$) and clusters 4 and 5 (p-value $= 0.0273$). (**a**) Optimal K plot. (**b**) Cluster scatter plot. (**c**) Survival curves

three covariates show distinct patterns in cluster 5, with the highest bilirubin, highest creatinine, and shortest cold ischemia time of the organ relative to all other clusters (Table 4).

Table 3 Distribution of number of alcoholic patients who received liver transplant (No. patients), number of patients who died after transplant (No. events), and the corresponding median survival time after transplant of patients (Median), and its 95% confidence interval (lower confidence limit, LCL, and upper confidence limit, UCL) with respect to spectral clusters

Cluster number	No. patients	No. events	Median	LCL	UCL
One	292	213(72.9)	2097	1817	2352
Two	121	78(64.5)	2026	1465	2184
Three	300	190(63.3)	2139	1822	2547
Four	201	132(65.7)	2197	1846	2552
Five	86	59(68.6)	1722	1149	2199
Total	1000	672			

Table 4 Median values of the most important covariates: bilirubin level (Bilirubin), creatinine level (Creatinine), and cold ischemia time for the organ (Cold_Isch) of alcoholic patients after receiving liver transplant, and the corresponding 95% confidence interval (lower confidence limit, LCL, and upper confidence limit, UCL) within each cluster formed by spectral clustering

Cluster number	Bilirubin	LCL	UCL	Creatinine	LCL	UCL	Cold_Isch	LCL	UCL
One	4.35	0.70	19.01	1.20	0.52	4.00	6.75	2.41	13.72
Two	3.45	0.83	16.95	1.20	0.59	4.00	7.00	2.23	13.40
Three	4.50	1.00	27.5	1.30	0.70	4.70	7.18	3.00	15.00
Four	3.50	0.60	20.30	1.00	0.46	4.10	7.00	1.00	14.00
Five	31.75	20.91	54.75	3.00	0.52	5.71	6.28	2.67	12.96

6.3 Clustering Using Persistent Homology in Dimension Zero (PH0)

Cluster identification using persistent homology incorporates Betti numbers obtained from the Vietoris–Rips filtration levels in PH0. Different filtration values yield different groups of connected components in PH0. In our data analysis, we used a threshold filtration value of 0.12 that provides a significantly large number of connected components or clusters in PH0. The threshold value that we chose in our data setting is greater than 0.041559, the upper limit for the $(1-\alpha)100\%$ confidence band. A point that forms a connected component is said to be significant if the death time is outside of the $(1-\alpha)100\%$ confidence band, Fig. 6a. If the death time is within the confidence band, the point is considered to be a noise point. From the persistence diagram in dimension zero with the corresponding 95% confidence band (Fig. 6a) and the persistent barcode (Fig. 6b), we can identify how many significant persistent features (clusters) in PH0 exist. From those persistent features, we chose the most significant ones (seven zero-dimensional features) at a filtration value of 0.12, a value which is outside of the confidence band, [0, 0.041559].

The clusters identified, using a filtration value of $\epsilon = 0.12$, the 1000 alcoholic patients who performed liver transplant and were distributed as 285 patients in cluster one, of which 63.9% died with a median survival time of 2160 days after liver transplant; 100 patients in cluster two, of which 61% died with a median survival

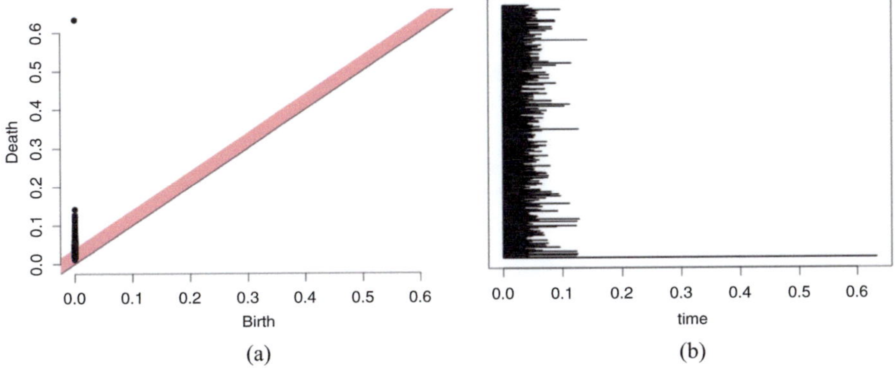

Fig. 6 (**a**) Persistence diagram and the corresponding 95% confidence band for the zero-dimensional features, clusters, at a Vietoris–Rips filtration value of $\epsilon = 0.12$. (**b**) Barcode representations of the most persistent zero-dimensional features (long-lived connected components) at a Vietoris–Rips filtration value of $\epsilon = 0.12$ for the liver transplant data. (**a**) Confidence band. (**b**) β_0-barcode

Table 5 Distribution of number of alcoholic patients who received liver transplant (No. patients), number of patients who died after transplant (No. events), and the corresponding median survival time after transplant of patients (Median), and its 95% confidence interval (lower confidence limit, LCL, and upper confidence limit, UCL) with respect to PH0 clusters

Cluster number	No. patients	No. events	Median	LCL	UCL
One	285	182(63.9)	2160	1832	2587
Two	100	61(61.0)	2508	2192	2949
Three	347	239(68.9)	1815	1674	2172
Four	86	58(67.4)	2486	2139	2916
Five	55	42(76.4)	1827	1108	2442
Six	90	61(67.8)	1473	1163	2540
Seven	37	29(78.4)	2180	1403	3448
Total	1000	672			

time of 2508 days after the transplant; and 347 patients in cluster three, of which 68.9% died at a median survival time of 1815, as shown in Table 5. This shows that cluster three is mostly formed by those patients who have lower survival time than those patients in clusters one and two.

An assessment of the difference in survival experience can be seen in Table 5, and suggests that the patients in the first cluster have a better median survival time compared to patients in clusters two and four. Compared to the remaining clusters, patients in the sixth cluster possess the lowest median survival time (1473 days); the cluster with the next lowest median survival time is cluster three at 1815 days, as shown in Fig. 7b.

When clustering using PH0, the seven clusters formed showed differences in the median values for the most important covariates included in the study, such as

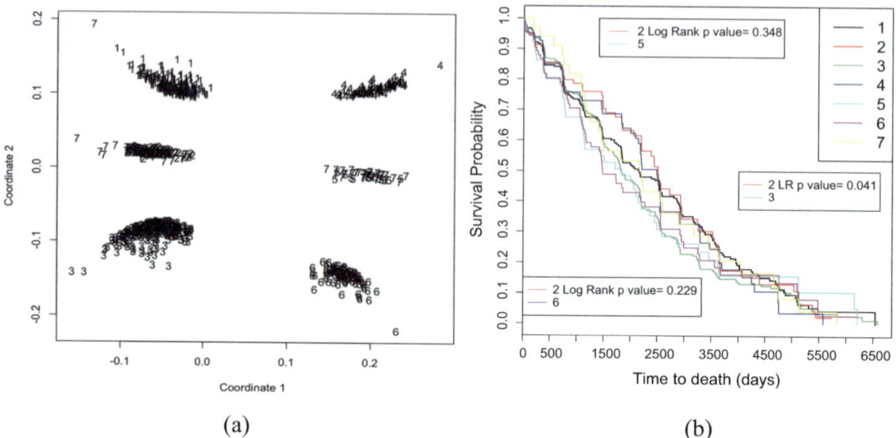

(a)

(b)

Fig. 7 (**a**) 2D multidimensional scaling scatter plot representation of the seven clusters formed by PH0 at a filtration value of $\epsilon = 0.12$ for alcoholic patients receiving liver transplant. (**b**) Survival curves for the corresponding seven patients clusters formed by PH0 at a filtration value of $\epsilon = 0.12$ and log-rank p-value comparison of cluster survival curves for alcoholic patients in clusters 2 and 3 (p-value = 0.041), clusters 2 and 5 (p-value = 0.348), and clusters 2 and 6 (p-value = 0.229). (**a**) 2D cluster representation. (**b**) Survival curves

Table 6 Median values of the most important covariates: bilirubin level (Bilirubin), creatinine level (Creatinine), and cold ischemia time for the organ (Cold_Isch) of alcoholic patients after receiving liver transplant, and the corresponding 95% confidence interval (lower confidence limit, LCL, and upper confidence limit, UCL) within each cluster formed by PH0

Cluster number	Bilirubin	LCL	UCL	Creatinine	LCL	UCL	Cold_Isch	LCL	UCL
One	3.70	0.80	32.25	1.30	0.60	4.27	7.00	3.00	14.94
Two	4.45	0.95	31.74	1.20	0.60	4.10	6.71	3.09	13.00
Three	5.00	0.80	43.37	1.27	0.50	4.71	7.00	2.20	13.64
Four	5.70	0.85	31.92	1.00	0.50	4.00	6.80	1.00	13.75
Five	3.30	0.44	37.56	1.10	0.54	4.00	6.16	2.83	16.46
Six	6.85	0.72	34.21	1.10	0.50	4.00	6.83	2.02	13.00
Seven	3.40	0.69	52.21	1.24	0.60	5.65	6.30	2.85	24.78

differences in median bilirubin. The highest, which is 6.85 (95% CI, 0.72, 34.21), is possessed by patients in cluster six and the smallest by patients in cluster five, at 3.30 (95% CI, 0.49, 37.56). As shown in Table 6, with respect to the covariate creatinine level of liver transplant patients, those patients in cluster four showed the lowest creatinine level of 1.00 (0.50, 4.00), and with regard to cold ischemia time of the organ (Cold_Isch), the lowest is found in those patients that form cluster five, 6.16 (95% CI, 2.83, 16.46).

The median survival time in PH0 clusters also varies, ranging from 1473 to 2508 (Table 5). Creatinine level is registered as lowest in the 4th PH0 cluster, in which the third lowest median cold ischemia time of the organ is also found (Table 6).

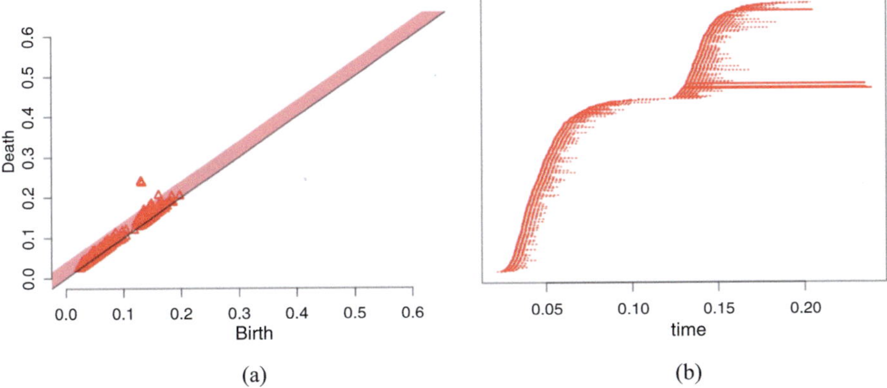

Fig. 8 (**a**) Persistence diagram and the corresponding 95% confidence band for the generated one-dimensional features (loops) from the Vietoris–Rips complexes. (**b**) Barcode representations of the most persistent one-dimensional features from the Vietoris–Rips complexes correspond to long-lived loops for the liver transplant data. (**a**) Confidence band. (**b**) β_1-barcode

6.4 Representative Loops Obtained via Persistent Homology in Dimension One (PH1)

We also apply persistent homology to further investigate whether there are some other features that can be explored in PH1. To do this, we considered 1000 alcoholic patients used in the PH0 analysis, as shown in Fig. 8. The figure shows that, in PH1, there are some features (loops) that persist longer than others. From the PH1 barcode representation Fig. 8b, we can identify that there are some significant features; about five features were found to be significant and to persist for a long period of time.

In this section, we wish to visualize and explore the most persistent components that form loops in PH1 homology formed when using a Vietoris–Rips filtration. The persistent generators of the first homology group arising from the Vietoris–Rips complexes correspond to long-lived loops formed by the group of alcoholic patients that received liver transplants. We remind readers that there can be many generators. In this text, "generator" refers to one of the many possible generators for the first homology group. We identify the five most persistent loops, as shown in Fig. 9, and it is clearly visible from the plots that not all alcoholic patients form the top most significant persistent loops. Those patients that have a significant contribution in forming the persistent loops and last longer as a component in these loops are displayed. About seven alcoholic patients were identified as generators of more than one persistent loop; however, for the analysis procedure we treated them explicitly as a generator for the most persistent one compared to the other, not considered as a member of more than one loop generator. As we can see from the persistence diagram, Fig. 8a, few loops were found to be significant.

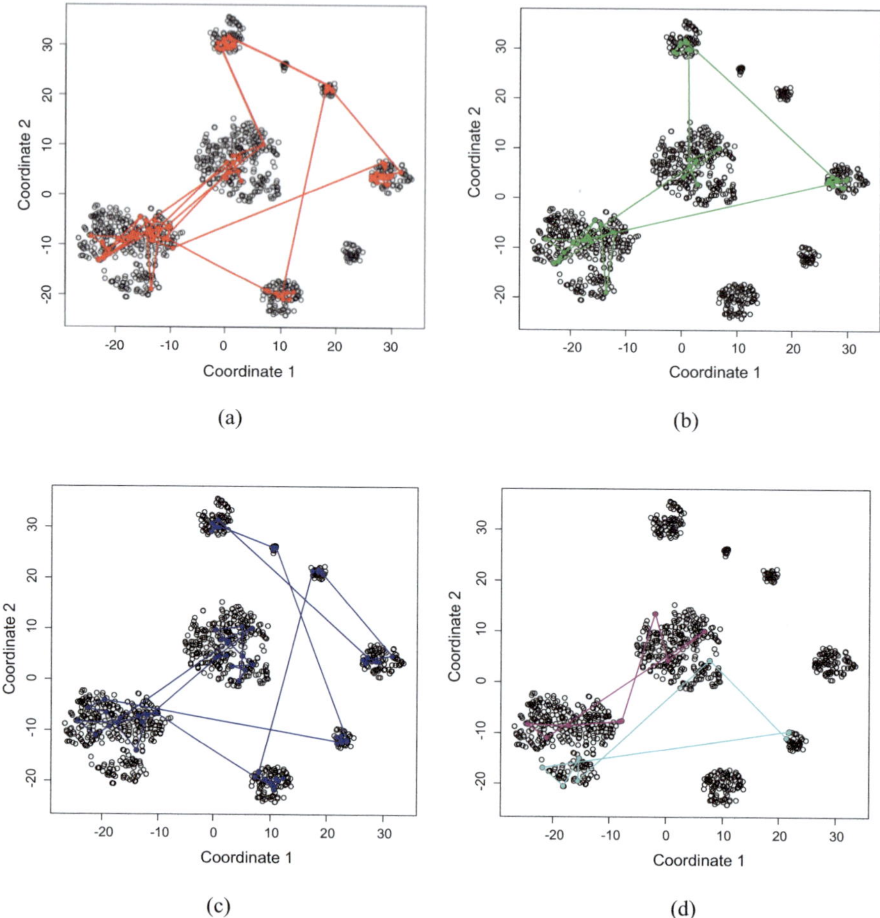

Fig. 9 2D multidimensional scaling scatter representation of the top five most persistent one-dimensional feature generators (we connect generators to form loops) from the Vietoris–Rips complexes that correspond to long-lived loops in PH1 for alcoholic patients receiving liver transplant data: first, second, third, fourth, fifth representative loops are represented by red, green, blue, blue-green, and purple, respectively. (**a**) First representative loop generators. (**b**) Second representative loop generators. (**c**) Third representative loop generators. (**d**) Fourth and fifth representative loop generators

After identifying those patients that form the first five most persistent loops, we perform an assessment for differences in survival experience between the patients forming these loops. We find that the second loop has the highest median survival time at 5312 days, and the next groups possessing lower median survival times are loops three and one with median survival times of 2852 and 2523 days, respectively. The persistent loops also revealed that about 113 alcoholic patients form persistent loop one and seven patients form loop four. From those patients that form persistent

Table 7 Distribution of number of alcoholic patients who received liver transplant (No. patients), number of patients who died after transplant (No. events), and the corresponding median survival time after transplant of patients (Median), and its 95% confidence interval (lower confidence limit, LCL, and upper confidence limit, UCL) with respect to PH1 persistent loops

Cluster number	No. patients	No. events	Median	LCL	UCL
One	113	55(48.7)	2523	2047	3524
Two	52	16(30.8)	5312	2635	NA
Three	75	34(45.3)	2852	2197	4393
Four	7	4(57.1)	844	389	NA
Five	8	2(25.0)	NA	2540	NA
Total	345	111			

loops one and four, about 48.7% in persistent loop four and about 57.1% of them in loop one died after receiving the transplant, as shown in Table 7. However, in persistent loops two and five, only 30.8% and 25% of them die after receiving transplant in each group, respectively.

In addition, these persistent components were evaluated based on the survival experiences of their constituent members, and as seen from Table 7 and from Fig. 10, showed that patients who formed persistent loop five had a much better survival experience even though they were few in number as compared to other patients that formed the remaining loops. Similarly, patients that formed persistent loops two and three showed an improved survival experience than those patients that formed loop four and one—both persistent loops that showed poor survival experience. Furthermore, survival experiences were analyzed for significant differences using log-rank test and it was found that patients that form persistent loop two showed better survival experiences than those patients that form the first loop (p-value = 0.0045), as shown in Fig. 10c. Even though patients that formed persistent loops four and five are too few in comparison to the other loops, patients in persistent loop four showed a poorer survival experience than patients in loop five (p-value = 0.0014), as seen from Fig. 10a.

Representative loops in PH1 were also assessed for trends in covariates as in other methods, and it was found that patients in cluster five have the highest median bilirubin level, 9.7 (95% CI, 1.58, 13.05) compared to the remaining clusters, but showed the highest median cold ischemia time of the organ (Cold_Isch), 9.91 (95% CI, 5.021, 44.950). Patients in cluster two are characterized as the second highest in bilirubin level and cold ischemia time of the organ, and the highest in creatinine level compared to the others. However, cluster three contains patients that have the lowest median bilirubin level. On the other hand, patients in cluster one showed the lowest median creatinine level compared to patients in other clusters (see Table 8).

The median survival times across PH1 representative loops vary widely: The 2nd loop expressed the highest median survival time of 5312 while the 4th loop showed the lowest median survival time of 844, as shown in Table 7. The second cluster with the highest survival times seems to be comprised of patients with the shortest ischemia time.

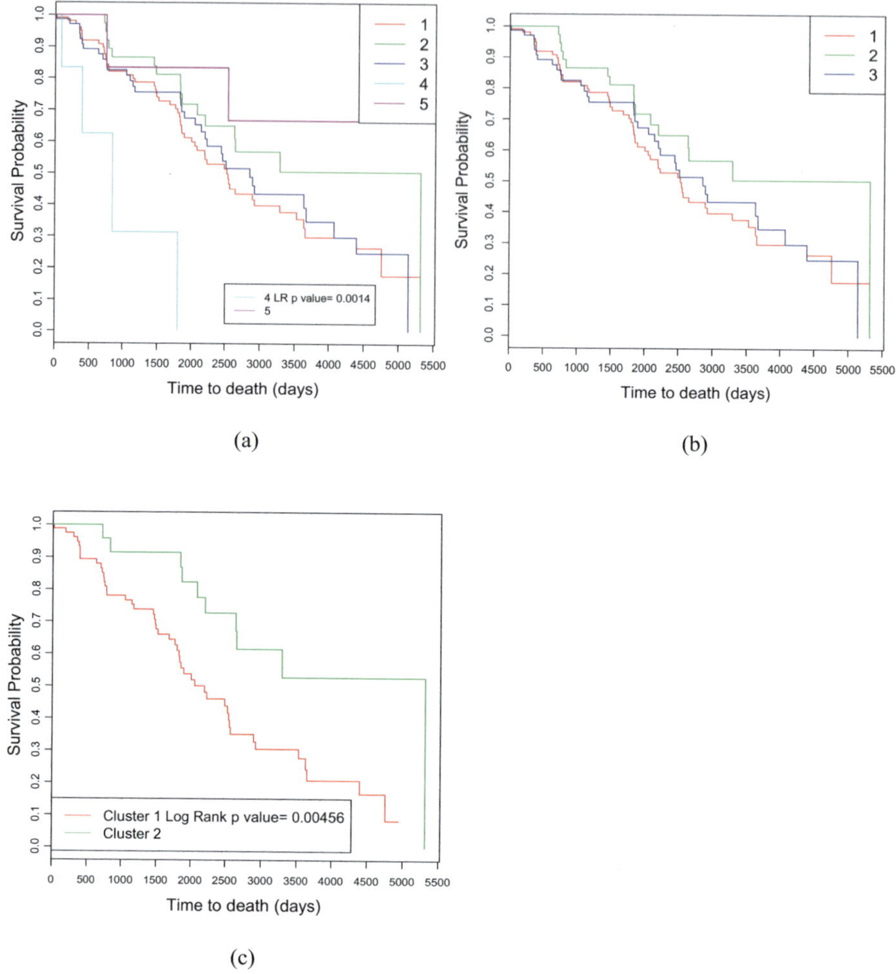

(a)

(b)

(c)

Fig. 10 Survival curves representation of the top five most persistent one-dimensional features generators from the Vietoris–Rips complexes that correspond to long-lived loops in PH1 for alcoholic patients receiving liver transplant: the first, second, third, fourth, fifth representative loop survival curves are represented by red, green, blue, blue-green, and purple, respectively. And corresponding log-rank p-value comparison for survival curves of loop generators for alcoholic patients data, clusters 4 and 5 (p-value = 0.0014), and clusters 1 and 2 (p-value = 0.0046). (**a**) Survival curves for five representative loops. (**b**) Survival curves of representative loops 1–3. (**c**) Survival curves for representative loops 1 and 2

We next compare clustering patterns for the SRTR-liver transplant data obtained by PAM, spectral clustering, and persistent homology. We denote each clustering method as PAM, Spectral, PH0, or PH1 when partitioning around medoids, spectral clustering, and persistent homology in dimensions zero and one, respectively.

Table 8 Median values of the most important covariates: bilirubin level (Bilirubin), creatinine level (Creatinine), and cold ischemia time for the organ, cooling time of the organ before transplant (Cold_Isch) of alcoholic patients after receiving liver transplant, and the corresponding 95% confidence interval (lower confidence limit, LCL, and upper confidence limit, UCL) within each loop formed by the representative cycles (generators) in PH1

Cluster number	Bilirubin	LCL	UCL	Creatinine	LCL	UCL	Cold_Isch	LCL	UCL
One	3.00	0.68	13.84	0.91	0.57	3.18	7.00	2.80	12.52
Two	4.65	1.30	16.80	0.94	0.50	4.30	6.98	2.07	14.00
Three	2.80	0.57	11.36	1.15	0.60	4.00	7.00	2.85	11.66
Four	8.4	1.52	64.29	4.00	1.76	4.00	8.60	3.62	9.30
Five	9.7	1.58	13.05	1.24	1.24	2.20	9.91	5.02	44.95

7   Comparisons of Clusters and Representative Loops Obtained via PAM, Spectral, PH0, and PH1

In this section, we compare and discuss clusters, of alcoholic patients who received liver transplant, resulted in applying PAM, spectral, PH0, and PH1 methods. According to the confidence interval for the persistence of persistent homology clusters, there are about 38 statistically significant clusters (see Fig. 6b) in Sect. 6.3. For simplicity, we inspect the first seven most statistically significant clusters.

When comparing clusters formed using partitioning around medoid (PAM) and persistent homology in dimension zero, as shown in Table 9, almost all clusters constitute an equal number of patients with the same patient partition in both clustering techniques. For example, the first PAM cluster consists of 286 patients of which about 285 are also members that form a cluster in PH0. Similarly, the remaining clusters also share the majority of common members in both clustering methods.

Considering the clusters formed by PH0 and spectral clustering, the optimum number of clusters that can be formed by spectral clustering is about five ways. These five clusters consist of combinations of patients that form different clusters in PH0 (see Table 10). For instance, cluster one in spectral clustering shares most of the patients that are classified as clusters one and three in PH0. Cluster three in spectral clustering is also formed from the majority of patients that form clusters one and three in PH0. Moreover, patients that form cluster four in spectral clustering come mostly from patients that form clusters four and six in PH0. A similar trend is observed in comparing the clusters formed by spectral clustering and PAM, as shown in Table 11.

For the representative loops formed by PH1, we only consider the top five loops. The representative loops formation is a combination of some of the clusters formed by both spectral and PAM clustering. As shown in Tables 12 and 13, for instance, the first representative loop formed using PH1 is a combination of most of the patients that form clusters three and four in spectral clustering, whereas this is a combination of PAM clusters three, four, and five. Again, for loop two in PH1, the representative

Table 9 Descriptive table for comparing the distribution of number of alcoholic patients who received liver transplant identified in each of the seven clusters formed by two methods: PH0 (columns) and PAM clustering (rows)

Cluster number	One	Two	Three	Four	Five	Six	Seven	Total
One	285	0	0	0	0	0	1	286
Two	0	100	0	0	0	0	1	101
Three	0	0	347	0	0	0	0	347
Four	0	0	0	86	5	0	0	91
Five	0	0	0	0	2	90	0	92
Six	0	0	0	0	44	0	2	46
Seven	0	0	0	0	4	0	33	37
Total	285	100	347	86	55	90	37	1000

Table 10 Descriptive table for comparing the distribution of number of alcoholic patients who received liver transplant identified in each of the seven clusters by PH0 (represented in columns) and five clusters identified by Spectral clustering (represented in rows)

Cluster number	One	Two	Three	Four	Five	Six	Seven	Total
One	93	31	125	11	20	10	2	292
Two	42	13	51	0	10	3	1	121
Three	104	37	106	19	10	15	9	300
Four	24	9	38	50	12	48	24	201
Five	22	11	27	6	3	14	1	86
Total	285	100	347	86	55	90	37	1000

Table 11 Descriptive table for comparing the distribution of number of alcoholic patients who received liver transplant identified in each of the seven clusters by PAM clustering (represented in columns) and five clusters identified by Spectral clustering (represented in rows)

Cluster number	One	Two	Three	Four	Five	Six	Seven	Total
One	93	31	125	13	11	17	2	292
Two	43	13	51	0	3	10	1	121
Three	104	37	106	21	15	8	9	300
Four	24	9	38	51	49	6	24	201
Five	22	11	27	6	14	5	1	86
Total	286	101	347	91	92	46	37	1000

elements come from patients that form clusters three and four in spectral clustering, and in PAM, it is from clusters three, four, and five. Few subjects in PAM clusters five, six, and seven belong to the fifth cluster of PH0 (See Table 9). Although the patients in each cluster obtained from the three clustering methods are not the same, the first five clusters obtained from PAM and PH0 are identical, with the exception of a few patients. The distributions across PAM and PH0 are similar, and we conclude that PAM and PH0 result in similar distribution of their clusters for this dataset.

Table 12 Descriptive table for comparing the distribution of number of alcoholic patients who received liver transplant identified in each of the representative loops formed by the persistent generators using PH1 (represented in columns) and seven clusters identified by PAM clustering (represented in rows)

Cluster number	One	Two	Three	Four	Five	Total
One	13	8	14	1	4	40
Two	11	0	11	0	0	22
Three	39	20	15	3	4	81
Four	20	12	9	0	0	41
Five	21	12	8	0	0	41
Six	0	0	11	3	0	14
Seven	9	0	7	0	0	16
Total	113	52	75	7	8	255

Table 13 Table for comparing the distribution of number of alcoholic patients who received liver transplant identified in each of the representative loops using PH1 (represented in columns) and five clusters identified by Spectral clustering (represented in rows)

Cluster number	One	Two	Three	Four	Five	Total
One	15	3	17	2	3	40
Two	0	0	2	0	0	2
Three	56	26	39	1	5	127
Four	42	23	17	1	0	83
Five	0	0	0	3	0	3
Total	113	52	75	7	8	255

8 Conclusions and Future Research

The data reported here have been supplied by the Minneapolis Medical Research Foundation (MMRF) as the contractor for the Scientific Registry of Transplant Recipients (SRTR). The interpretation and reporting of these data are the responsibility of the author(s) and in no way should be seen as an official policy of or interpretation by the SRTR or the US Government.

We carried out RSF with seven predictor variables: gender, age, blood type, levels of creatinine, bilirubin, and albumin, and cold ischemia time of the organ. RSF identified the five covariates most important in predicting survival time after transplant to be creatinine, cold ischemia time of the organ, bilirubin, albumin, and height. First, we focused on the three most important variables in interpreting the statistically significant clusters obtained from the three clustering methods. The clusters determined by PH0 are identical to those determined by PAM, except for few patients. On the other hand, significant differences were found between clusters formed by PH0 and spectral clustering as well as between those formed by PAM and spectral clustering.

Table 14 Clusters with the highest survival time for each clustering method. Minimum values among 1000 patients of survival time, bilirubin, creatinine, and cold ischemia time are 0, 0.20, 0.20, and 0, respectively

Clustering method	Median survival time	Number of failure	Median bilirubin	Median creatinine	Median ischemia time
Spectral	2197	132(201)	3.50	1.00	7.00
PAM	2508	62(101)	4.40	1.20	6.80
PH0	2508	61(100)	4.45	1.20	6.71
PH1	5312	16(52)	4.65	0.94	6.98

Table 15 The clusters with the lowest survival time for each clustering method. Maximum values among 1000 patients of survival time, bilirubin, creatinine, and cold ischemia time are 6583, 66.7, 8.80, and 46, respectively

Clustering method	Median survival time	Number of failure	Median bilirubin	Median creatinine	Median ischemia time
Spectral	1722	59(86)	31.75	3.00	6.28
PAM	1473	61(92)	6.85	1.10	6.95
PH0	1473	61(90)	6.85	1.10	6.83
PH1	844	4(7)	8.40	4.00	8.60

The median survival times in clusters formed by PH0 and PAM are similar, while the loops formed by PH1 are better discriminated with respect to their survival times. Comparing the loops constructed by PH1, the second loop consists of patients with the longest survival times and the fourth loop consists of patients with the least survival times. The cooling time of the organ for patients in the second loop is the least, while patients in the fourth loop have the most (ignoring the fifth loop with the highest cold ischemia time as it contains only two failed patients).

In the below tables, we summarize patterns of clusters in terms of minimal and maximal survival times (in days) in each clustering technique. PH1 loops seem effective in detecting patients with extreme survival times. The loops formed by PH1 are able to identify patients with the lowest median survival time of 844 days, and groups of patients with the highest median survival time of 5312 days. Those patients who survived longer tended to have a lower level of bilirubin and creatinine, as well as short cold ischemia time. The PH1 representative loops with a survival time of 844 consists of patients with higher values of all three these variables (Tables 14 and 15).

It is interesting to note that our findings based on clustering analyses are concordant with already-published experience in liver transplantation. The cold ischemia time has extensively been associated with worse survival [13]. High bilirubin and high creatinine are also known factors associated with worse survival experience after liver transplantation [8, 27].

Since the two predictor variables, albumin and height, are more or less uniformly distributed over all the clusters, we do not interpret them; however, we present

Table 16 Median and corresponding 95% confidence interval (LCL and UCL) of the covariates within each cluster formed by PH0

Clusters	Albumin	LCI	UCI	Height	LCL	UCL
One	3.0	1.80	4.59	177.80	162.57	190.50
Two	3.1	1.70	4.15	177.80	162.27	195.50
Three	2.9	1.70	4.20	177.80	162.36	193.04
Four	3.1	2.11	4.39	165.05	152.40	175.30
Five	3.0	1.77	4.52	175.26	152.94	195.44
Six	3.0	1.82	4.50	162.58	152.45	175.26
Seven	3.0	1.52	4.02	165.10	152.37	183.20

Table 17 Median and corresponding 95% confidence interval (LCL and UCL) of the covariates within each cluster formed by PAM

Clusters	Albumin	LCI	UCI	Height	LCL	UCL
One	3.0	1.80	4.59	177.80	162.74	190.50
Two	3.1	1.70	4.15	177.80	162.28	195.50
Three	2.9	1.70	4.20	177.80	162.36	193.04
Four	3.1	2.13	4.38	165.10	152.40	175.29
Five	3.0	1.83	4.50	162.58	150.55	175.26
Six	2.9	1.66	4.64	175.63	160.02	197.15
Seven	3.0	1.52	3.93	165.00	150.37	182.88

Table 18 Median and corresponding 95% confidence interval (LCL and UCL) of the covariates within each cluster formed by Spectral Clustering

Clusters	Albumin	LCI	UCI	Height	LCL	UCL
One	2.9	1.80	4.2	177.80	162.57	193.04
Two	2.8	1.70	4.1	182.88	170.18	195.58
Three	3.0	1.85	4.4	175.26	160.01	187.96
Four	3.0	1.60	4.5	165.00	152.40	178.00
Five	3.2	1.46	5.2	172.72	152.40	185.00

relevant tables in the appendix (Tables 16, 17 and 18). As we were only able to apply clustering techniques to 1000 data points, we plan to reanalyze the entire data set in the future.

The authors [28] encountered difficulty in finding relatively small clusters using PAM and k-means clustering when one or more larger clusters are present. We find, in our study, that PH1 seems to be sensitive enough to detect small patient "clusters": The fourth loop with seven patients and the fifth loop with 8 patients are both examples of this. It may be a coincidence that these loops reveal certain patterns that are not detected by other clustering techniques, and thus future simulation studies with several real-world complex data sets that compare different clustering methods including RSF and support vector clustering are necessary. Based on what we have illustrated in Fig. 2 and what we have observed in our survival data, we speculate that PH1 might be useful in detecting clusters composed of a small number of elements. However, this conjecture and whether k-dimensional holes in general can be considered high-dimensional clusters still remains to be theoretically verified.

Acknowledgements We would like to thank Matthew Pietrosanu for his support in the computation process. We also thank Jisu Kim and Dmitriy Morozov for their insightful discussions and suggestions. We would like to acknowledge funding support provided by McIntyre Memorial Fund and NSERC DG 293180.

Appendix

Here we describe RSF and its algorithm in more detail. Random forest is an ensemble method that combines a number of trees by taking the same number of bootstrap samples from the original data and growing a tree on each bootstrap sample [3, 17]. Random Survival Forests is an extension of random forest for the analysis of right-censored survival data. As is well known, constructing ensembles from base learners, such as trees, can significantly improve learning performance [19, 22]. Random Survival Forests (RSF) is modeled closely after Breiman's approach. Random Survival Forests naturally inherit many of their good properties from the RF. It is user-friendly and fairly robust. Individual trees in a random survival forest are not pruned and used for decisions: Instead predictions are based on all trees grown from the bootstrap samples. Moreover, RSF is data-driven, with a derivation based on data and free of model assumptions, unlike other standard methods used for the analysis of survival data [22]. RSF is also known for its consistency in that the survival function converges uniformly to the true population survival trend [20, 23].

Random Survival Forest Algorithm

The algorithm used for a random survival forest is similar to the algorithm used in random forest and is adapted for survival data. To fill this need, the R package randomForestSRC is available for implementing random forests for survival, regression, and classification [19–22]. The algorithm used by randomForestSRC for survival data is broadly described as follows and shown in Fig. 3.

1. Draw B bootstrap samples from the original data, B = number of trees, (*ntree*).
2. Grow a tree for each bootstrapped data set. At each node of the tree, randomly select predictors (covariates) for splitting on *mtry*. Split on a predictor using a survival splitting criterion. A node is split on the predictor maximizing survival differences across daughter nodes.
3. Grow the tree to full size under the constraint that a terminal node should have no less than node size unique events (deaths).
4. Calculate an ensemble cumulative hazard estimate by combining information from the *ntree* trees. One estimate for each individual in the data is calculated.

5. Compute an out-of-bag (OOB) error rate for the ensemble derived using the first b trees, where $b = 1, \ldots, ntree$.

Assume we are at node h of a tree during its growth and that we seek to split h into two daughter nodes. We introduce some notation to help discuss how the various splitting rules work to determine the best split. Assume that within h there are n individuals. Denote their survival times and 0–1 censoring information by, $(T_1, \delta_1), \ldots, (T_n, \delta_n)$. An individual l is said to be right censored at time T_l if $\delta_l = 0$, otherwise the individual is said to have died at T_l if $\delta_l = 1$. In the case of death, T_l will be referred to as an event time (death time). An individual l who is known to have been alive at T_l but whose exact time of death is unknown is called right-censored.

A proposed split at node h on a given predictor x is always of the form $x \leq c$ and $x > c$. Such a split forms two daughter nodes (a left and a right daughter) and two new sets of survival data. A good split maximizes survival differences across the two sets of data. Let $t_1 < t_2 < \ldots < t_N$ be distinct death times in the parent node h, and let d_{ij} and Y_{ij} equal the number of deaths and number of individuals at risk at time t_i in the daughter nodes $j = 1, 2$. Note that Y_{ij} is the number of individuals in daughter node j who are alive at time t_i or who have an event (death) at time t_i [16, 19, 22]. More precisely,

$$Y_{i1} = \#\{l : T_l \geq t_i, x_l \leq c\}, \ Y_{i2} = \#\{l : T_l \geq t_i, x_l > c\} \tag{4}$$

where x_l is the value of x for individual $l = 1, 2, \ldots, n$. Finally, define $Y_i = Y_{i1} + Y_{i2}$ and $d_i = d_{i1} + d_{i2}$. Let n_j to be the total number of observations in the daughter j, thus, $n = n_1 + n_2$. Note that $n_1 = \#\{l : x_l \leq c\}$ and $n_2 = \#\{l : x_l > c\}$.

Ensemble Estimation

The randomForestSRC package produces an ensemble estimate for the cumulative hazard function. The cumulative hazard function is the predictor and main input for the computation of performance error in random survival forest. The ensemble estimation is derived as follows. First, for each tree grown from a bootstrap data set, we estimate the cumulative hazard function for the tree. This is accomplished by grouping hazard estimates by terminal nodes. Consider a specific node h. Let $\{t_{ih}\}$ be the distinct death times in h and let d_{ih} and Y_{ih} equal the number of deaths and individuals at risk at time t_{ih}. The cumulative hazard estimate for node h is defined as

$$\hat{H}_h(t) = \sum_{t_{ih} \leq t} \frac{d_{ih}}{Y_{ih}}. \tag{5}$$

Each tree provides a sequence of such estimates, $\hat{H}_h(t)$. If there are M terminal nodes in the tree, then there are M such estimates. To compute $\hat{H}(t|x_l)$ for an individual l with predictor x_l, simply drop x_l down the tree, and then the terminal node for l yields the desired estimator [19, 22]. More precisely,

$$\hat{H}(t|x_l) = \hat{H}_h(t), \; if \;\; x_l \epsilon h. \tag{6}$$

Note this value is computed for all individuals l in the data. This estimate (6) is based on one tree. To produce our ensemble, we average (6) over all $ntree$ trees. Let $\hat{H}_h(t|x_l)$ denote the cumulative hazard estimate (6) for tree $b = 1, \ldots, ntree$. Define $I_{l,b} = 1$ if l is an OOB point for b, otherwise set $I_{l,b} = 0$. The OOB ensemble cumulative hazard estimator for l is [22]

$$\hat{H}_e^*(t|x_l) = \frac{\sum_{b=1}^{ntree} I_{l,b} \hat{H}_b(t|x_l)}{\sum_{b=1}^{ntree} I_{l,b}}. \tag{7}$$

Prediction Error

To compute the error rate, we need to have an OOB ensemble estimator $\hat{H}_e^*(t|x)$ and then, using this estimator to fine the Harrell's concordance index, we can measure the performance of the survival predictions by taking into account the censoring of subjects [19, 22]. Before computing a concordance index, we must define what constitutes a worse predicted outcome. Let $t_1^*, \ldots, t_N^*$ denote all unique event times in the data. Individual i is said to have a worse predicted survival experience than j if

$$\sum_{k=1}^{N} \hat{H}_e^*\left(t_k^*|x_i\right) > \sum_{k=1}^{N} \hat{H}_e^*\left(t_k^*|x_j\right). \tag{8}$$

The prediction error in using RSF for the overall grown trees is found to be 33.26% and, from the out of bag (OOB) error rate, we can see that as the number of random trees grown in a forest increases, the OOB error slowly stabilizes and becomes closer to the mentioned overall error rate (see Fig. 11c). Hence, we can say that the forest grown based on the 7154 alcoholic patients has a good predictive ability of the survival experience (or the cumulative death hazard) after liver transplant based on the behaviors and characteristics of a new patient receiving liver transplant.

In a random survival forest, after identifying and ranking the importance of variables or a similar procedure for variable selection in some other statistical models, we can apply those variables for predicting the survival probability at different times (see Fig. 11a and Table 19). As shown in Fig. 11d, we plotted the survival probability curves of the first five alcoholic patients who received liver transplant. From the plot, we can see how the survival experience of patients considered in the study behaves.

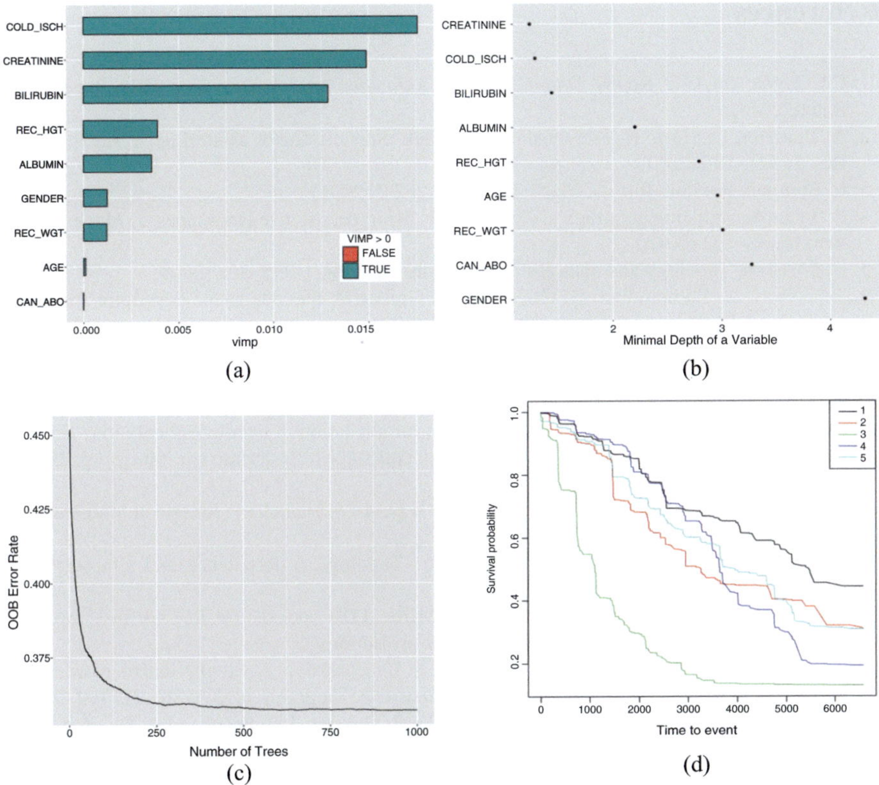

Fig. 11 (**a**) Variable importance plot, (**b**) variable depth plot for RSF, (**c**) OOB error rate, and (**d**) Predicted survival curves for the first five patients based on RSF. (**a**) VIMP plot. (**b**) Depth plot. (**c**) OOB error plot. (**d**) Survival curves

Table 19 Variable Importance (VIMP) of patient characteristics using Random Survival Forest (RSF) for liver transplants in alcoholic patient data

Variable	Depth	Depth rank	VIMP rank	VIMP value
CREATININE	1.242	1	2	1.489e-02
COLD_ISCH	1.292	2	1	1.759-02
BILIRUBIN	1.447	3	3	1.286e-02
ALBUMIN	2.206	4	5	3.566e-03
HEIGHT	2.793	5	4	3.883e-03
AGE	2.962	6	8	1.115e-04
WEIGHT	3.008	7	7	1.221e-03
ABO	3.273	8	9	−2.199e-05
GENDER	4.321	9	8	1.246e-03

References

1. C.C. Aggarwal, C.K. Reddy, *Data Clustering. Algorithms and Applications* (CRC Press, Boca Raton, 2014)
2. A. Ben-Hur, D. Horn, H. Siegelmann, V. Vapnik, Support vector clustering. J. Mach. Learn. Res. **2**, 125–137 (2001)
3. L. Breiman, Random forests. Mach. Learn. **45**, 5–32 (2001)
4. P. Bubenik, Statistical topological data analysis using persistence landscapes. J. Mach. Learn. Res. **16**, 77–102 (2015)
5. G. Carlsson, Topology and data. Bull. Am. Math. Soc. **46**(2), 255–308 (2009)
6. A. Collins, A. Zomorodian, G. Carlsson, L.J. Guibas, A barcode shape descriptor for curve point cloud data. Comput. Graph. **28**, 881–894 (2004)
7. V. de Silva, G. Carlsson, Topological Approximation by Small Simplicial Complexes. Eurographics Symposium on Point-Based Graphics (2004).
8. P. Dutkowski, C.E. Oberkofler, K. Slankamenac, M.A. Puhan, E. Schadde, B. Mullhaupt, A. Geier, P.A. Clavien, Are there better guidelines for allocation in liver transplantation? A novel score targeting justice and utility in the model for end-stage liver disease era. Ann. Surg. **254**(5), 745–753 (2011)
9. H. Edelsbrunner, J. Harer, *Computational Topology: An Introduction* (American Mathematical Society, Providence, 2010)
10. H. Edelsbrunner, D. Letscher, A. Zomorodian, Topological persistence and simplification. Discrete Comput. Geom. **28**, 511–533 (2002)
11. B.T. Fasy, J. Kim, F. Lecci, C. Maria, V. Rouvreau. TDA package for R. *Statistical tools for topological data analysis*, (2014), https://cran.r-project.org/.
12. B.T. Fasy, F. Lecci, A. Rinaldo, L. Wasserman, S. Balakrishnan, A. Singh, Statistical inference for persistent homology: confidence sets for persistence diagrams. arXiv:1303.7117v2 (2013)
13. S. Feng, N.P. Goodrich, J.L. Bragg-Gresham, D.M. Dykstra, J.D. Punch, M.A. DebRoy, S.M. Greenstein, R.M. Merion, Characteristics associated with liver graft failure: the concept of a donor risk index. Am. J. Transplant. **6**(4), 783–790 (2006)
14. J. Fridlyand, "Resampling Methods for Variable Selection and Classification: Applications to Genomics," Ph.D. thesis, University of California, Berkeley, Dept. of Statistics, (2001).
15. R. Ghrist, Barcodes: the persistent topology of data. Bull. Am. Math. Soc. **45**(1), 61–75 (2008)
16. L. Gilles, W. Louis, S. Antonio, G. Pierre, Understanding variable importance in forests of randomized trees, in *NIPS'13 Proceedings of the 26th International Conference on Neural Information Processing Systems* (Curran Associates Inc., Red Hook, 2013), pp. 431–439
17. T. Hastie, R. Tibshirani, J.H. Friedman, *The Elements of Statistical Learning: Data Mining, Inference and Prediction*, 2nd edn. (Springer, New York, 2009)
18. G. Heo, J. Gamble, P. Kim, Topological analysis of variance and the maxillary complex. J. Acoust. Soc. Am. **107**, 477–492 (2012)
19. H. Ishwaran, U.B. Kogalur, Random survival forests for r. R News **7**(2), 25–31 (2007)
20. H. Ishwaran, The effect of splitting on random forests. Mach. Learn. **99**, 75–118 (2015)
21. H. Ishwaran, U.B. Kogalur, *randomForestSRC: Random Forests for Survival, Regression and Classification (RF-SRC).* R package version 2.2.0 (2016). http://cran.r-project.org
22. H. Ishwaran, U.B. Kogalur, E.H. Blackstone, M.S. Lauer, Random survival forests. Ann. Appl. Stat. **2**, 841–860 (2008)
23. H. Ishwaran, U.B. Kogalur, X. Chen, A.J. Minn, Random survival forests for high-dimensional data. Stat. Anal. Data Min. **4**, 115–132 (2011)
24. L. Kaufman, P.J. Rousseeuw, *Finding Groups in Data: An Introduction to Cluster Analysis* (Wiley, New York, 2005)
25. D. Morozov, Dionysus: a C++ library for computing persistent homology (2007). http://www.mrzv.org/software/dionysus
26. H.S. Park, C.H. Jun, A simple and fast algorithm for k-medoids clustering. Expert Syst. Appl. **36**, 3336–3341 (2009)

27. M.S. Roberts, D.C. Angus, C.L. Bryce, Z. Valenta, L. Weissfeld, Survival after liver transplantation in the United States: a disease-specific analysis of the UNOS database. Liver Transpl. **10(7)**, 886–897 (2004)
28. M.J. van der Laan, K.S. Pollard, J. Bryan, A new partitioning round medoids algorithm. J. Stat. Comput. Simul. **73(8)**, 575–584 (2003)
29. U. von Luxburg, A tutorial on spectral clustering. Stat. Comput. **17**(4), 395–416 (2007)
30. K. Xia, G.W. Wei, Persistent homology analysis of protein structure, flexibility, and folding. Int. J. Numer. Methods Biomed. Eng. **30**(8), 814–844 (2014)
31. A. Zomorodian, G. Carlsson, Computing persistent homology. Discrete Comput. Geom. **33**, 249–274 (2005)

Pseudo-Multidimensional Persistence
and Its Applications

**Catalina Betancourt, Mathieu Chalifour, Rachel Neville, Matthew Pietrosanu,
Mimi Tsuruga, Isabel Darcy, and Giseon Heo**

Abstract While one-dimensional persistent homology can be an effective way to
discriminate data, it has limitations. Multidimensional persistent homology is a
technique amenable to data naturally described by more than a single parameter, and
is able to encoding more robust information about the structure of the data. However,
as indicated by Carlsson and Zomorodian (Discrete Comput Geom 42(1):71–271,
2009), no perfect higher-dimensional analogue of the one-dimensional persistence
barcode exists for higher-dimensional filtrations. Xia and Wei (J Comput Chem
36:1502–1520, 2015) propose computing one-dimensional Betti number functions
at various values of a second parameter and stacking these functions for each homological dimension. The aim of this visualization is to increase the discriminatory
power of current one-dimensional persistence techniques, especially for datasets
that have features more readily captured by a combination of two parameters. We
apply this practical approach to three datasets, relating to (1) craniofacial shape
and (2) Lissajous knots, both using parameters for scale and curvature; and (3)
the Kuramoto–Sivashinsky partial differential equation, using parameters for both
scale and time. This new approach is able to differentiate between topologically

C. Betancourt · I. Darcy
Department of Mathematics, The University of Iowa, Iowa City, IA, USA
e-mail: catalina-betancourt@uiowa.edu

M. Chalifour · M. Pietrosanu
Department of Mathematical & Statistical Sciences, University of Alberta, Edmonton, Alberta,
Canada
e-mail: mrc@ualberta.ca; pietrosa@ualberta.ca

R. Neville
Department of Mathematics, University of Arizona, Tucson, AZ, USA
e-mail: raneville@math.arizona.edu

M. Tsuruga
Department of Mathematics, University of California, Davis, Davis, CA, USA
e-mail: mtsuruga@math.ucdavis.edu

G. Heo (✉)
School of Dentistry, University of Alberta, Edmonton, Alberta, Canada
e-mail: gheo@ualberta.ca

© The Author(s) and the Association for Women in Mathematics 2018
E. W. Chambers et al. (eds.), *Research in Computational Topology*, Association
for Women in Mathematics Series 13, https://doi.org/10.1007/978-3-319-89593-2_10

equivalent geometric objects and offers insight into the study of the Kuramoto–Sivashinsky partial differential equation and Lissajous knots. We were unable to obtain meaningful results, however, in our applications to the screening of anomalous facial structures, although our method seems sensitive enough to identify patients at severe risk of a sleep disorder associated closely with craniofacial structure. This approach, though still in its infancy, presents new insights and avenues for the analysis of data with complex structure.

1   Introduction and Motivation

Persistent homology has emerged as a primary tool in topological data analysis, encoding both topological and geometric characteristics of data. This is done in the following way: a filtration of topological spaces is associated to data. Using standard methods from algebra and topology, the topological features are observed through the filtration and encoded in the invariant called a barcode. The most common filtration is built on point cloud data by forming a simplicial complex, on points within some proximity parameter, ϵ. Changes in homology of the simplicial complex are observed as the proximity parameter increases, which captures multiscale topological features. The reader is directed to [4, 12, 14] for a careful introduction to persistent homology.

Although persistence barcodes are useful for distinguishing many objects, there are limitations. For example, persistent homology alone cannot distinguish between data sampled from a circle or from an ellipse of similar size, as both shapes exhibit a topological hole. The discriminating power of persistent homology could benefit by exploiting multiple characteristics of the data. Collins et al. [6] suggested constructing simplicial complexes for a fixed proximity parameter ϵ by varying a curvature threshold parameter κ, approximating the curvature numerically using each point's nearest neighbors. The resulting filtration would start with a simplicial complex determined by the fixed value of ϵ and one would observe the changing homology of this complex as points with low curvature are filtered out of consideration. However, this method requires a choice of proximity parameter to be made initially. The removal of such choices is a strength of persistent homology.

Multidimensional persistent homology provides the mathematical framework to characterize datasets naturally described by more than one parameter. For instance, we may wish to study homological changes in protein unfolding patterns as a function of both time and scale, or even more simply perform several different filtrations on geometric objects incorporating scale, curvature, or torsion. Figure 1 shows the evolution of a simplicial complex under changes in both scale and curvature parameters ϵ and κ, respectively.

It has long been proven [5], however, that no perfect higher-dimensional analogue of the one-dimensional persistence barcode—that is, a complete discrete algebraic invariant—exists for higher-dimensional filtrations. The problem of generalizing a barcode from a single dimension to multiple dimensions remains a difficult

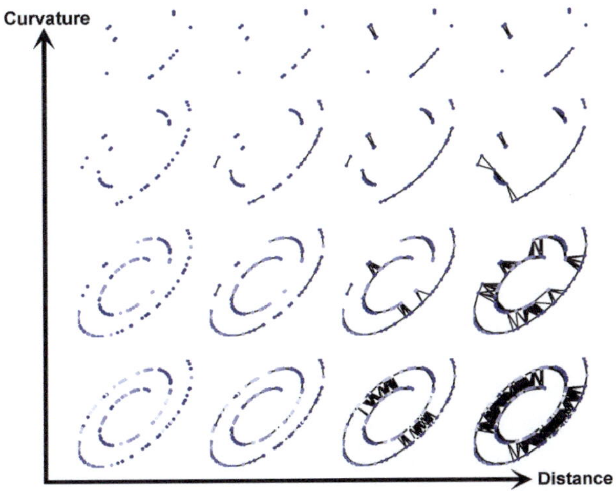

Fig. 1 Visualization of a bifiltration parametrized by curvature κ and distance, or scale ϵ for a simulated dataset sampled from an elliptical spiral. As the curvature parameter κ increases on the vertical axis, points with curvature less than κ are removed from the dataset. The scale parameter, ϵ, increases along the horizontal axis. At a pair of filtration values, ϵ_i and κ_j, a Vietoris–Rips (VR) complex is built on points in the dataset within a distance of ϵ_i, after removing points from the data with curvature less than κ_j (Edges, but not faces of the VR complex are shown)

mathematical problem [18]. The theoretical limitation of accessible representations of multidimensional persistence, however, does not stop researchers' heuristic applications to a wide range of fields including brain networks [17], Vicsek and D'Orsogna models [26], and biomolecular data [29], for example.

In this article, we consider a practical approach to approximating multidimensional persistent homology. Looking to the 2015 work of Xia and Wei [29], we further develop their methods of *pseudo-multidimensional persistence*. We are particularly motivated by the task of increasing the discriminatory power of current one-dimensional persistence techniques. To accomplish this goal, we use the pseudo-multidimensional persistence technique to incorporate additional parameters into the standard one-variable filtration and a computationally-feasible framework that admits the desired discriminatory power.

We close this introduction with a brief description of the following sections. Section 2 proposes a nonparametric method to compare the point-cloud datasets using pseudo-multidimensional persistence in two variables, which we apply to the simple problem of distinguishing between a circle and an ellipse using scale and curvature parameters. In Sects. 3–5, we apply methods for pseudo-multidimensional persistence to real-world and simulated datasets, including craniofacial shape, Lissajous knots, and the Kuramoto–Sivashinsky partial differential equation. We conclude this work in Sect. 6 with comments on these heatmap methods and suggestions for future work.

2 Illustration of Heatmap Method with Circle and Ellipse

We return to the challenge of distinguishing a circle from an ellipse to illustrate the extension of one-dimensional persistence to a two-dimensional heatmap built from filtrations of both scale and curvature. In general, an appropriate method for the calculation of curvature at each point of a point-cloud dataset may differ depending on the nature of the data. In this paper, for noisy data obtained in applied settings, such as the craniofacial data of Sect. 3, we employ a generalization of the hyper circle-fitting algorithm [2] to arbitrary dimensions to estimate curvature: The curvature at a given point will be defined as the reciprocal of the radius of the hypersphere fit locally to the dataset at that point. Our choice of algorithm is discussed further in Sect. 3. In pure settings, such as with the Lissajous knots of Sect. 4, it may be possible to calculate curvature analytically.

Consider a set of points sampled from a uniform random distribution on the boundary of a circle of radius 1 and the boundary of an ellipse with major axis 2 and minor axis 1, both with a small amount of added noise. We mix noise into the sample data in order to replicate real data that is generally composed of true signal which we wish to estimate, and random noise that we wish to ignore. We first set a dimension p as the dimension of the homology to examine, and further fix finite sequences $(\kappa_i)_{i=1}^{K}$ and $(\epsilon_j)_{j=1}^{J}$ of *curvature* and *distance thresholds*, respectively.

For each dataset and for each choice of curvature and distance thresholds κ_i and ϵ_j, we calculate the p-th Betti number, β_p at the specified thresholds. More specifically, only considering those points of the dataset with estimated curvature at least the chosen threshold κ_i, we calculate the number $\beta_p(\kappa_i, \epsilon_j)$ of p-th dimensional homological components at the scale parameter ϵ_j. In implementation, this is done by constructing a persistence barcode for the Vietoris–Rips filtration on each dataset after removing points with estimated curvature less than κ_i, and computing the number of p-th dimensional homological components at each distance threshold ε_j. The heatmap matrix formed by associating $\beta_p(\kappa_i, \epsilon_j)$ to each position (i, j). As shown in Fig. 2, the heatmaps for a circle and an ellipse visually differentiate between them. In particular, the heatmap representing the first homological dimension for a circle shows a topological hole of fairly uniform prominence (visualized as the values across a single row of the heatmap). This continues up to a certain curvature threshold, above which there is no longer a hole, as expected. There is a notable decrease in the prominence of the topological hole for higher curvature thresholds of the ellipse.

Motivated by the above result, we now propose a nonparametric test of dissimilarity between the heatmaps generated from two sample spaces S_A and S_B. To do this, we compare the heatmap derived from a single sample of S_B with heatmaps obtained from N samples of S_A. For our example, S_A and S_B represent, respectively, the sample space of points sampled from the circle and ellipse described above.

1. Fix a sequence of curvature and scale thresholds $(\kappa_i)_{i=1}^{K}$ and $(\epsilon_j)_{j=1}^{J}$, respectively. For $n = 1, 2, \cdots, N$, create the $K \times J$ *heatmap matrix* H_n^A, where each

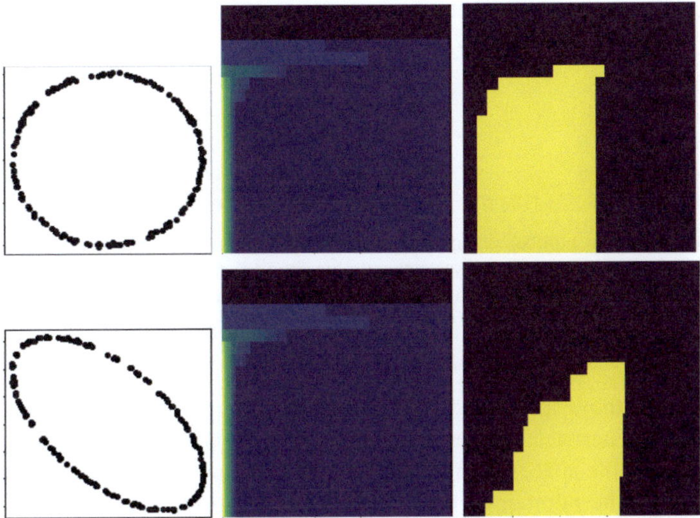

Fig. 2 (Top row) β_0 (center) and β_1 (right) heatmap plots of the circle on the left. (Bottom row) β_0 (center) and β_1 (right) heatmap plot of the ellipse on the left. The horizontal and vertical axes of each heatmap indicates values of the scale and curvature parameters ϵ and κ, respectively. In all heatmaps above, Betti numbers β are plotted as $\ln(\beta + 1)$, so that heatmap colors show better contrast for large differences in Betti number. We can see, visually, the differences in the circle and ellipse heatmaps for each dimension

of these heatmaps is generated from a sample of S_A. In general, we define the entry in the i-th row and j-th column of a heatmap matrix to be β_p at curvature and scale threshold parameters κ_i and ϵ_j, respectively, divided (or "normalized") by the total number of points in the sample used to obtain the heatmap.

2. Calculate the mean heatmap matrix $\overline{H}^A$ of the heatmap matrices obtained in the previous step. More precisely, $\overline{H}^A = \frac{1}{N} \sum_{n=1}^{N} H_n^A$.

3. For $n = 1, 2, \cdots, N$, calculate the dissimilarity d_n between each heatmap H_n^A and the mean heatmap $\overline{H}^A$: As an example, we could define such dissimilarity by

$$d_n = d(H_n^A, \overline{H}^A) = \sum_{i=1}^{K} \sum_{j=1}^{E} \left[\ln\left((H_n^A)_{ij} + 1\right) - \ln\left((\overline{H}^A)_{ij} + 1\right) \right]^2.$$

4. Calculate the dissimilarity d_{obs} between the heatmaps $\overline{H}^A$ and H^B, where H^B is the heatmap generated from a point-sampling of S_B.

5. (Step 4 may be iterated using repeated samples obtained from S_B, if available). One may take the p-value for this test to be the proportion of iterations of Step 4 where $\{d_n > d_{obs}\}$. A smaller p-value indicates stronger evidence that the heatmaps generated from each of the sample spaces S_A and S_B are different.

Table 1 Parameters and results for the proposed non-parametric test of dissimilarity for comparing point-cloud datasets

Test #	N	n_A	n_B	K	J	s	p_0	p_1
1 (circle, ellipse)	20	200	200	20	100	0.3	0	0
2 (ellipses)	20	200	200	20	100	0.3	1	1
3 (3-leaf, 4-leaf clovers)	20	250	250	20	50	0.3	0	0
4 (wedge product of two spheres)	20	600	600	20	100	0.3	0	0

The columns represent number of bootstrap samples (N), number of points in each point-cloud (n_A, n_B), number of curvature and scale thresholds used (K and J, respectively), radius of the hypersphere used for curvature estimation (s), and p-values for the proposed test in dimension 0 (p_0) and 1 (p_1)

To test the above procedure, we performed a number of experiments with generated data. These tests are summarized in Table 1, where the S_A and S_B columns describe the sample spaces being compared, N the number of samples drawn from S_A, n_A and n_B the number of points sampled from each sample space, K and J the number of curvature and scale parameter threshold values chosen, s the radius of the neighborhood used during curvature estimation, and p_0 and p_1 the p-values calculated by the test when comparing the zero- and one-dimensional homology groups of S_A and S_B.

Curvature thresholds are chosen to be $\{q_{k/K} | k = 0, \ldots, K - 1\}$, where q_p is the p-th quantile for all computed curvature values, for $p \in [0, 1]$. Fifty scale thresholds were similarly chosen using quantiles of the death times of dimension-0 homological components obtained from univariate persistent homology applied to each of datasets under consideration. The sample spaces compared in each test, S_A and S_B, respectively, are as follows. Test #1 compares an ellipse with vertical axis of length 2 and horizontal axis of length 1, rotated 45 degrees CCW against the unit circle. Test #2 compares an ellipse with vertical axis of length 2 and horizontal axis of length 1, rotated 45 degrees counterclockwise against the same ellipse but rotated 45 degrees clockwise. Test #3 compares a "three-leaf clover" shape with a "four-leaf clover" shape, as shown in Fig. 3. Finally, Test #4 tests out method in three dimensions, and compares the wedge product of two spheres with the same radius of 1 against the wedge product of two spheres, one with radius 1 and the other with radius 2.

From the results of our tests, we see that the above approach is sensitive enough to distinguish between shapes that are homologically equivalent but have different curvatures, as discussed at the beginning of this section. Furthermore, this method is not overly sensitive to falsely distinguish between different point-samples of the same object transformed under rotation. The proposed technique is clearly reliant on the curvature estimates obtained for each point of the sampled dataset, particularly when the sampled data contains a moderate amount of noise. While the hyper circle-fitting algorithm is superior to numerous other circle-fitting methods—such as the Kasa, Pratt, or Taubin fits—in terms of essential bias [2], we note that the curvatures obtained are sensitive to moderate levels of noise in the sampled data. This is

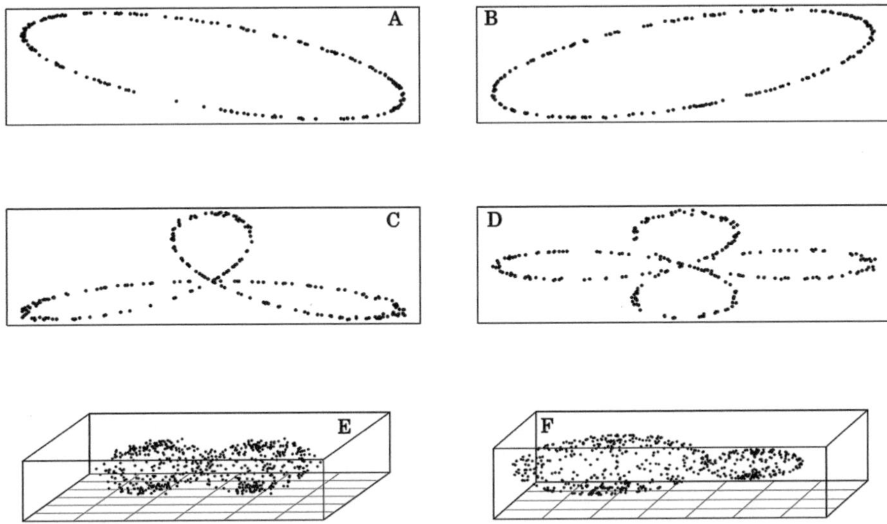

Fig. 3 (**a** and **b**) Test #2 datasets sampled from two ellipses identical up to rotation. (**c** and **d**) Test #3 dataset sampled from curves resembling a three- and four-leaf clover. (**e** and **f**) Test #4 dataset sampled from the wedge product of two spheres: only in one dataset are the two spheres of the same radius

particularly true when the size of the neighborhood used to locally fit a hypersphere to the given data (and, by extension, obtain the estimated curvature as the inverse of the radius of this hypersphere) is small, or when the number of points used to fit the hypersphere is otherwise small.

The remainder of this paper will detail several applications of heatmaps and is organized as follows. The first task is to predict a patient's risk of developing a specific sleep disorder from facial scans. Next, we distinguish specific types of knots. Third, we show how heatmaps may be used to distinguish between dynamic data of pattern evolution, driven by different parameter values. In these applications, some knowledge about the data and possible distinguishing characteristics drives the choice of the types of filtrations (for instance, curvature and distance in the previous example). It is our hope that through these examples, we demonstrate the added utility of the heatmap representation of a pseudo-multidimensional persistence in a variety of contexts.

3   Craniofacial Shape Analysis

We examine the use of heatmaps in an applied setting, namely, in the analysis of point-cloud scans obtained from pediatric patients for the screening of obstructive sleep apnea (OSA). Pediatric OSA is a sleep disorder with serious health problem that may lead to high blood pressure, behavioral challenges, or altered overall growth. The gold standard for pediatric OSA diagnosis is overnight polysomnography in a hospital or sleep clinic [19]. In many countries, however, access to overnight polysomnography is severely limited and many children are unable to obtain a proper diagnosis before treatment. This absence of an accessible diagnostic method thus prompts the search for alternative screening methods. Evidence continues to demonstrate a link between craniofacial shape and pediatric OSA [13]. As part of a larger research initiative examining alternative OSA screening methods leaning on orthodontic expertise and statistical techniques for shape and high-dimensional data analysis, we examine the utility of the proposed heatmap method for identifying children at risk of developing OSA on the basis of 3D craniofacial scans.

The dataset used in our investigation is composed of 3D facial scans obtained from 31 children 2–17 years of age recruited from the Stollery Children's Hospital at the University of Alberta. Prior ethics approval has been granted by the University of Alberta's Research Ethics Board. All of the recruited subjects have undertaken overnight polysomnography and have had 3D photos of their face taken. Based on polysomnography results, an apnea–hypopnea index (AHI) for each subject was calculated. These AHI measurements obtained from polysomnography are commonly used to classify patients according to OSA risk severity into one of four categories: no likely risk ($AHI < 1$), mild risk ($1 \leq AHI < 5$), moderate risk ($5 \leq AHI < 10$), and severe risk ($AHI \geq 10$). In this preliminary analysis, we only seek to classify patients into one of two derived groups: no/mild risk ($AHI \leq 5$) and moderate/high risk ($AHI > 5$).

The 3D photo of each patient's face is itself a point-cloud in $\mathbb{R}^3$. For each point, a measure of curvature is computed (see Fig. 4). This task is more difficult than when working with analytic curves, such as the Lissajous knots in the following section. To approach the issue of curvature estimation in this applied setting, we seek to locally fit a sphere in the neighborhood of each point of the dataset. In general, we define the curvature of a dataset at a given point to be the reciprocal of the radius of a hypersphere fit to the dataset in a neighborhood of that point. In our case, we take this neighborhood to be a cube with a manually-selected side length centered at the point whose curvature is being estimated. The relative size of this neighborhood is shown in Fig. 5 for reference.

For the estimation of point-cloud curvature, we implement Al-Sharadqah and Chernov's hyper circle-fitting algorithm [2] in three dimensions. We choose to use this particular method due to its simplicity in generalizing to fit hyperspheres to data of arbitrary dimension and in its property of being essentially unbiased. In other words, while the hyperspheres fit by this method are not unbiased, the level of bias is proportional to the reciprocal of the number of points being fit to. As a result, we have control over the amount of bias in the parameters of the fitted hypersphere—an improvement over other estimation methods [2]. To create the heatmaps for a filtration in both scale and curvature, we must first choose a set of scale and curvature thresholds. In our case, we select curvature thresholds to be the quantiles of the curvature values computed for all points in all point-clouds from

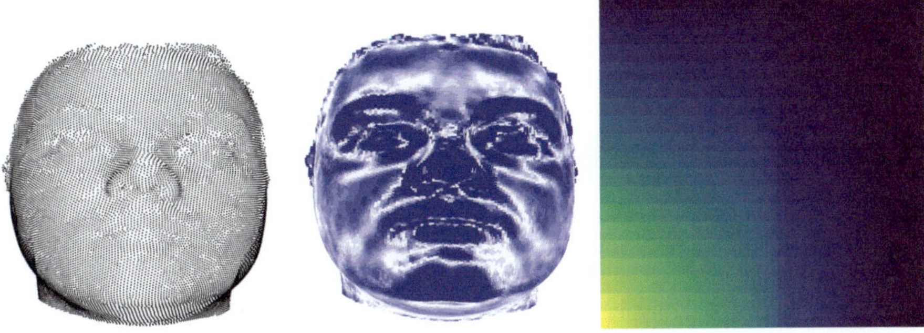

Fig. 4 (Left) An example of a raw point-cloud dataset. (Middle) The same point-cloud dataset as on the left, with points colored according to curvature. White represents low estimated curvature, while deep blue represents high estimated curvature. (Right) The β_0 heatmap corresponds to the two-parameter filtration of the dataset on the left by both scale and curvature

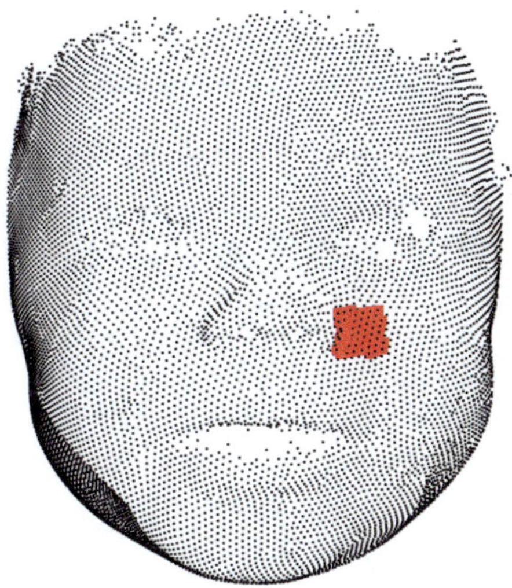

Fig. 5 A point-cloud extracted from a single patient. Shown in red is the neighborhood used to estimate the curvature of some point in the centre of the red-shaded region

all patients. Twenty quantiles were chosen, namely, $\{q_{k/19}|k = 0, \ldots, 19\}$, where q_p is the p-th quantile for all computed curvature values, for $p \in [0, 1]$. Fifty scale thresholds were similarly chosen using quantiles of the death times of dimension-0 homological components obtained from univariate persistent homology applied to each of the patient datasets. We stress that the existence of an "optimal" method of choosing these scale and curvature threshold values has not been investigated in any depth and remains a topic for future work.

Having obtained heatmaps for each patient, we now proceed to address the task of classifying patients into the no/low risk or the moderate/severe risk groups. Due to the study design, a number of the pediatric patients are designated as controls with little to no risk of OSA: Using the heatmaps obtained from these

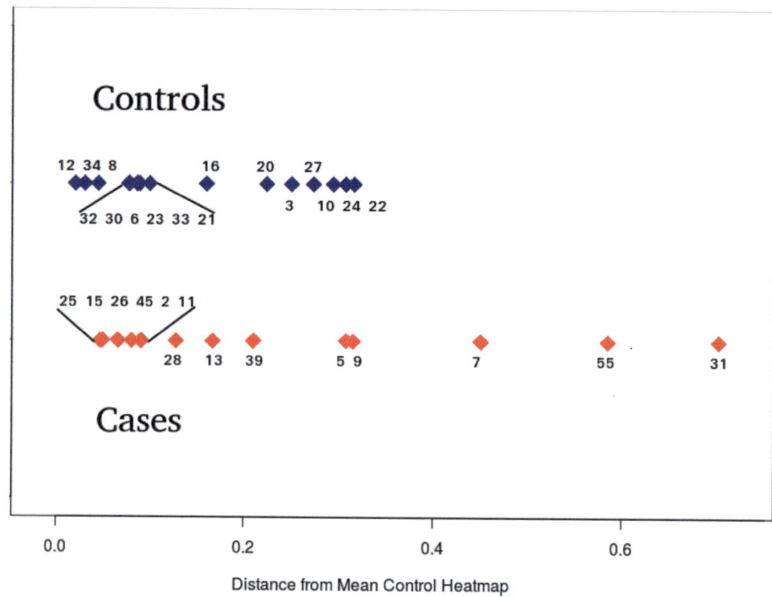

Fig. 6 A visual summary of the dissimilarity (x-axis) between each patient's computed heatmap and the average control heatmap. Controls are represented with blue points, while cases (that is, non-controls) are represented using red points. One case patient (numbered as patient #1), at a distance of 1.2 from the mean control heatmap, is not shown in this diagram to preserve scale. Clearly, the heatmap distance does not adequately distinguish cases from controls. Seven out of twelve cases were clustered with the most of control subjects. Subject #1 indicates a possible outlier; however, its actual AHI is 10.0, the 7th highest AHI. Subject #55 has the highest AHI of 40.2

control patients, we compute an average heatmap for the controls. From here, we calculate the dissimilarity between each patient's heatmap and the average control heatmap, according to the dissimilarity measure defined in the previous section. These dissimilarities, for both controls and cases (non-controls), are summarized in Fig. 6 and in Table 2, sorted by increasing AHI score.

It is immediately apparent that the distance from the mean control heatmap does not correlate well with AHI score. Indeed, patient 25, for example, has a heatmap closer to the mean control heatmap than nearly all control patients but has one of the highest AHI scores. Most case subjects (11 out of 15) were clustered together with control subjects. There are four cases (subjects 7, 55, 31, and 1 (not shown in the diagram)) that appear different from other subjects, however. Further investigating these subjects, we learn that subjects 1, 7, and 55 either have undergone or have scheduled a tonsillectomy and/or adenoidectomy surgery, suggesting that these patients have severe OSA symptoms. Subject 31 was recommended to an ears, nose, and throat specialist for further diagnosis.

Although our heatmap method is able to detect severe OSA cases, these results suggest that the approach used may not be appropriate in developing craniofacial

Table 2 Case information for the 31 children considered, sorted from left to right by increasing AHI score

ID	23	16	22	24	20	3	6	21	8	10	34	27	32	30	12	33
AHI	0.4	0.6	1.2	1.4	1.5	1.6	1.6	1.6	1.8	1.9	2.1	2.2	2.7	3.7	4	4.3
Cases	No	No	No	No	No	No	No	No	No	No	No	No	No	No	No	No
OSA	1	1	2	2	2	2	2	2	2	2	2	2	2	2	2	2
ID	11	15	26	9	31	39	5	13	1	7	2	25	45	28	55	
AHI	5	5.4	5.6	6.7	7.5	7.6	8.8	9.4	10	10.4	11.7	11.7	13.6	13.8	40.2	
Cases	Yes	Yes	Yes	Yes	Yes	Yes	Yes	Yes	Yes	Yes	Yes	Yes	Yes	Yes	Yes	
OSA	2	3	3	3	3	3	3	3	3	4	4	4	4	4	4	

The rows represent, respectively, patient ID, apnea-hyponea index (AHI), identification as a case or control, and specific OSA severity classification (no(1), mild(2), moderate(3), and severe(4))

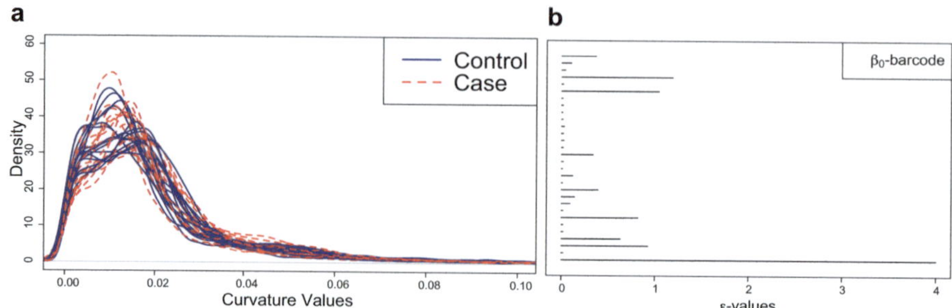

Fig. 7 (**a**) Estimated density function for the computed curvature values for each patient. Curves corresponding to control subjects are shown in blue, while those corresponding to cases are shown in red. (**b**) β_0-barcode based on the Kullback–Leibler measures between the density curves. β_0-barcode indicates six clusters among 31 subjects

shape analysis as a proxy for overnight polysomnography and the derived AHI scores. More generally, our results here leave much in the way of future work in this area, although demonstrated an ability to identify extreme OSA cases. It may be necessary to focus on a specific part of the face rather than the face in its entirely, or perhaps curvature and scale thresholds need to be selected in a way that will allow the resulting heatmaps to better highlight differences between the faces under study. Furthermore, it may be of use to apply different statistical techniques, such as those in machine learning, to this problem of heatmap classification.

As a means of comparison, we also apply one-dimensional persistent homology to craniofacial data. We estimate the probability density function of the curvature values computed for each patient (Fig. 7). We calculate Kullback–Leibler distance which measures the difference between two probability distributions over the same variable [16]. A β_0-barcode based on the Kullback–Leibler divergence between densities is shown in Fig. 7. Subjects corresponding to the six most persistent clusters are presented in Fig. 8 applying multidimensional scaling. Multidimensional scaling (MDS) is a nonlinear dimension reduction technique in which the data is assigned coordinates in a lower dimensional space in a way that most closely preserves the distance between points.

The largest cluster is formed with 17 subjects—8 cases and 9 controls. We also see a similar pattern in the second largest component composed of 3 cases and 2 controls. Univariate persistent homology is unable to differentiate between these two groups, while the heatmap approach seems to be able to do so for extreme cases. This result demonstrates that some benefits exist in choosing to use heatmaps over univariate persistent homology, particularly for the analysis of datasets described by multiple parameters.

Both the proposed heatmap and univariate persistent homology methods based on curvature fail to differentiate between obstructive sleep apnea patients and controls, except that bifiltration heatmap was capable of detecting a few extreme cases (see Fig. 6). We can think of three potential reasons for this: (1) curvature may not be

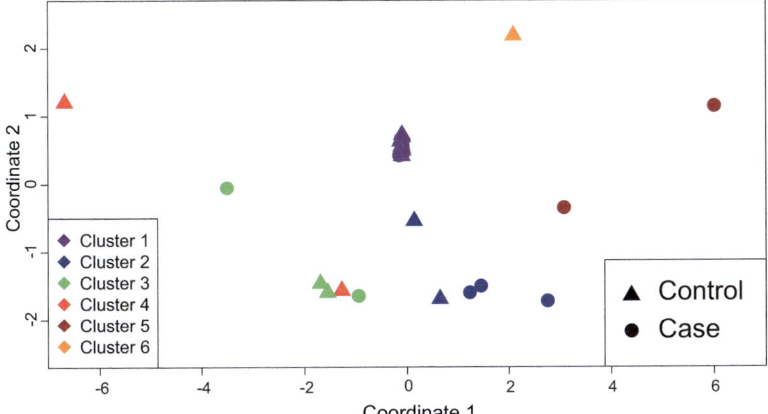

Fig. 8 Clustering patterns of subjects presented on first two MDS coordinates (coordinate 1 on X-axis and coordinate 2 on Y-axis). The subgroups (clusters) do not seem to relate to any common traits among the subjects. The largest cluster (in purple) consists of 17 subjects—8 cases and 9 controls. The second largest cluster (in blue) consists of five subjects—3 cases and 2 controls

the best way to measure craniofacial form, (2) our proposed heatmap method is not effective for complex data such as face shape, or (3) AHI may not be the most effective variable in determining the presence or absence of OSA as several researchers have suggested [30].

4 Lissajous Knots

In this section, we apply the heatmaps to the study of a geometric object, namely, Lissajous knots. The filtrations throughout this section use scale and curvature parameters. Our goal is to use known curves, Lissajous knots, to investigate how heatmaps can help us understand features of this data. In particular, as shown below, we deduce that the two flares seen in the heatmaps from these Lissajous knots correspond with the fact that the data points of highest curvature are separated into two disjoint sets. In the future, we would like to use this information to interpret the heatmaps of data sets from other one-dimensional curves (such as other types of knots and some time series data) where the density of data points is higher around curves.

A Lissajous knot [3] is a closed curve that is isotopic to some curve with parameterization of the form

$$x(t) = \cos(n_x t + \phi_x)$$

$$y(t) = \cos(n_y t + \phi_y)$$

$$z(t) = \cos(n_z t + \phi_z),$$

Fig. 9 The $(3, 4, 7, 0.1, 0.7)$ Lissajous knot given by the parameterization $(x(t), y(t), z(t)) = (\cos(3t + 0.1), \cos(4t + 0.7), \cos(7t))$ is of type 8_{21} according to Rolfsen's Knot Table [22]. The knot is colored by z value (vertical axis). We immediately observe that the curvature of this knot is largest at its highest and lowest heights, that is, where minimum and maximum values of z are attained. Visually, we see that the knot's curvature oscillates periodically between large and small values as the path of the knot is traced out

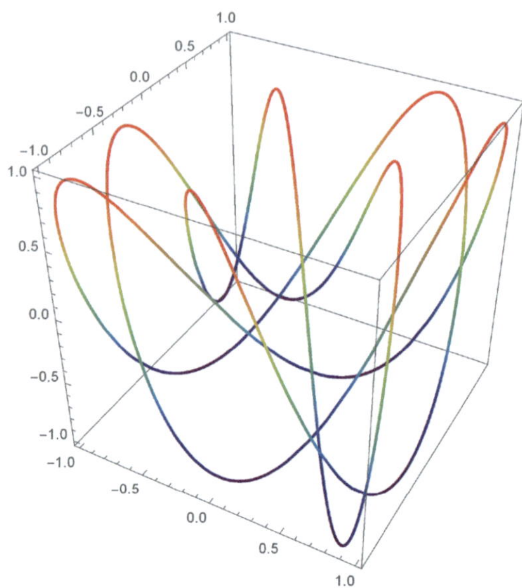

where $0 \leq t \leq 2\pi$ and i and j are any distinct elements of $\{x, y, z\}$. We require the frequencies n_i to be pairwise relatively prime integers, ϕ_i a real number, and $n_i \phi_j - n_j \phi_i$ not a multiple of π. These conditions ensure that the resulting curve is closed and without self-intersection. To simplify we may, with a change of variable, assume that $\phi_z = 0$: this parameterization places the Lissajous knot inside the cube centered at the origin with edge length 2. As such, we may denote a Lissajous knot by $(n_x, n_y, n_z, \phi_x, \phi_y)$. For example, Fig. 9 shows the $(3, 4, 7, 0.1, 0.7)$ Lissajous knot, given by $(x(t), y(t), z(t)) = (\cos(3t + 0.1), \cos(4t + 0.7), \cos(7t))$. Under Rolfsen's Knot Table classification [22], this knot is of type 8_{21}. According to Table 1 of [3], a given knot type may be associated with multiple values of $(n_x, n_y, n_z, \phi_x, \phi_y)$: As such, in this section, we restrict ourselves to exactly one parameterization for each of the knot types considered.

Lissajous knots have practical applications, particularly in the modelling of DNA [3], and have thus been well-studied. The knots in this family are highly symmetric, with curvature values changing predictably over a wide interval as a function of t. These properties make Lissajous knots an appropriate object of interest in our study of heatmaps where we may naturally incorporate both scale and curvature filtration parameters.

For each Lissajous knot considered, we create a dataset with size proportional to the knot's total arc length. The $(3, 4, 7, 0.1, 0.7)$ Lissajous knot, for example, with the parameterization given previously, has an arc length of 36.8. We thus choose to create the dataset for this knot using 368 sampled points. Since the knot itself is parameterized by t, we create the dataset S by picking 368 equally-spaced t values between 0 and 2π, that is, $S = \{t_k = 2\pi k/368 \mid k = 0, \ldots, 367\}$ and find their spatial coordinates $(x(t_k), y(t_k), z(t_k))$, for $k = 0, \ldots, 367$.

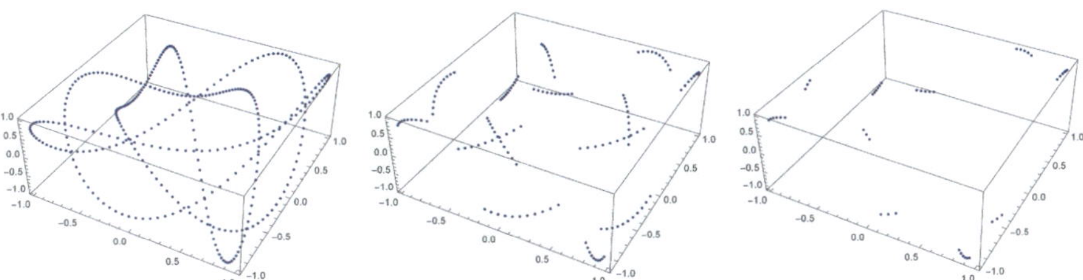

Fig. 10 Super-level set point-clouds from the (3, 4, 7, 0.1, 0.7) Lissajous knot dataset. (Left) All points with curvature greater than 0, that is, the entire dataset of 368 points. (Middle) All points with curvature greater than 1, consisting of 147 points. (Right) All points with curvature greater than 3, comprised of 47 points

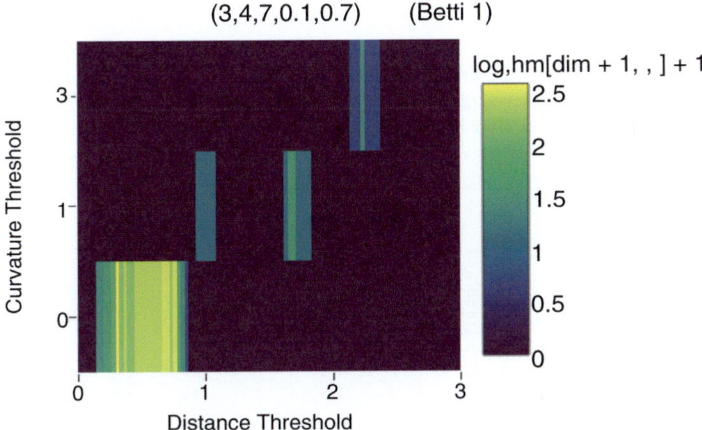

Fig. 11 Persistence heatmap for the first homology group of (3, 4, 7, 0.1, 0.7) Lissajous knot dataset. The one-dimensional heatmap for the entire dataset of 368 points is shown in the bottom row of the heatmap, with the one-dimensional heatmaps for the smaller datasets stacked on top in increasing order of curvature threshold used. The distance threshold increases along the x-axis from 0 to 3 using step size 0.3. The color scale on the right represents the natural log of the dimension-1 Betti numbers after being increased by 1

Using Mathematica, we then (analytically) compute the curvature, $\kappa(t)$, at each selected point, and create super-level sets $S_h = \{(x(t), y(t), z(t)) \in \mathbb{R}^3 : \kappa(t) > h\}$, for any real h, referred to as the curvature threshold. In short, S_h consists of all points of the dataset with curvature strictly greater than h.

Figure 10 shows three such super-level sets from the (3, 4, 7, 0.1, 0.7) Lissajous knot dataset. Figure 11 displays the persistence heatmap for the first homology group for this knot under various scale and curvature threshold values.

We can create a larger heatmap by choosing more curvature thresholds as in Fig. 12, which shows another heatmap for the (3, 4, 7, 0.1, 0.7) Lissajous knot. Along the x-axis is the scale parameter ranging from 0 to 3 with step size 0.3. Along the y-axis is the curvature parameter ranging from 0 to the maximum curvature by step size 0.1. The maximum curvature for each knot is different: For example, the we use maximum curvature 16.5 for the (3, 4, 7, 0.1, 0.7) knot.

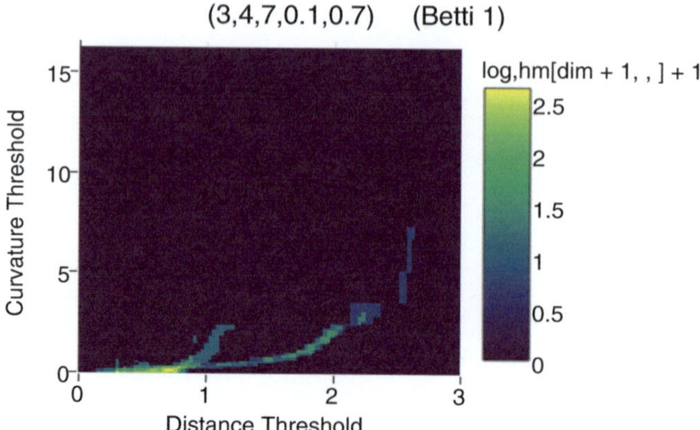

Fig. 12 Heatmap for the (3,4,7,0.1,0.7) Lissajous knot. The distance threshold increases along the *x*-axis from 0 to 3 by step size 0.3. The curvature threshold increases along the *y*-axis from 0 to its maximum curvature 16.5 by step size 0.1. The color scale on the right represents the natural log of the dimension-1 Betti numbers after being increased by 1

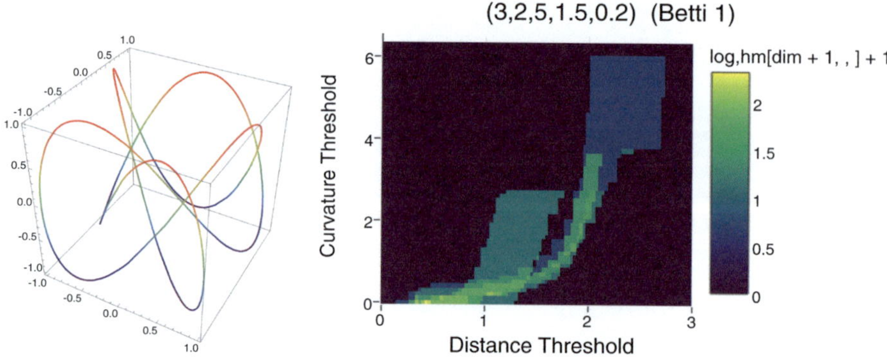

Fig. 13 Left: the (3, 2, 5, 1.5, 0.2) Lissajous knot with parametrization $(x, y, z) = (\cos(3t + 1.5), \cos(2t + 0.2), \cos(5t))$. Right: the heatmap for this knot. The distance threshold increases along the *x*-axis from 0 to 3 by step size 0.3. The curvature threshold increases along the *y*-axis from 0 to a maximum curvature 6.5 by step size 0.1. The color scale on the right represents the natural log of the dimension-1 Betti numbers after being increased by 1

Figures 13 and 14 show two more heatmaps for other Lissajous knots, namely, (3, 2, 5, 1.5, 0.2) and (2, 3, 11, 0.2, 0.7), respectively, along with their knot graphs. The (3, 2, 5, 1.5, 0.2) Lissajous knot is given by $(x, y, z) = (\cos(3t + 1.5), \cos(2t + 0.2), \cos(5t))$ and has total arc length of 26.4. We then choose to construct the dataset for this knot using 264 points. According to Rolfsen's Knot Table [22], this knot is of type 6_1. The (2, 3, 11, 0.2, 0.7) Lissajous knot is parameterized via $(x, y, z) = (\cos(2t + 0.2), \cos(3t + 0.7), \cos(11t))$ and has arc length of 47.8. We thus create the dataset for this knot using 478 points.

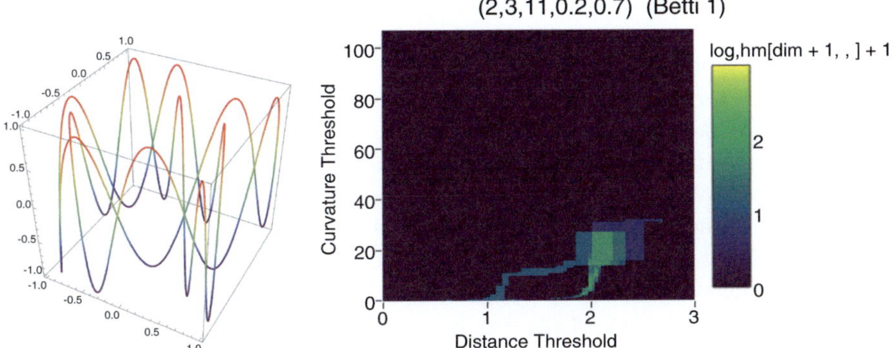

Fig. 14 Left: the (2,3,11,0.2,0.7) Lissajous knot with parametrization $(x, y, z) = (\cos(2t + 0.2), \cos(3t + 0.7), \cos(11t))$. (Right) Heatmap for this Lissajous knot. The distance threshold increases along the x-axis from 0 to 3 by step size 0.3. The curvature threshold increases along the y-axis from 0 to its maximum value 107 by step size 0.1

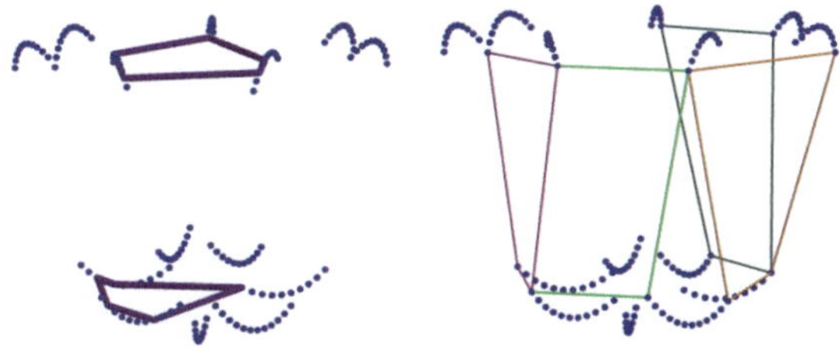

Fig. 15 Cycles in the super level set at curvature threshold 1 of the (3, 4, 7, 0.1, 0.7) Lissajous knot. (Left) Two cycles are present in the super-level set at distance threshold 1. (Right) Four cycles are present in the super-level set at distance threshold 1.7

A curiosity of Figs. 12, 13, and 14 is the two blue-green flares across each of the heatmaps. We investigate this further using the software ShortLoop[10] by visualizing the cycles in the first homology of our dataset. Figure 15 shows the cycles appearing in the super-level set of curvature threshold 1 for the (3, 4, 7, 0.1, 0.7) Lissajous knot at two different ε distance thresholds. We see two cycles at $\varepsilon = 1$ on the left, and four cycles at $\varepsilon = 1.7$ on the right.

As mentioned previously, Lissajous knots attain maximum curvature at their highest and lowest heights (z values). As a result, filtering our data by curvature may also be viewed as filtering by the absolute z-coordinate. For a given curvature threshold, our point-cloud can then, loosely speaking, be clustered into groups with a relatively high z and those with relatively low z. The flares in the heatmap represent these two clusters. When ε is smaller than the distance between the clusters, cycles in each cluster result, but no cycles exist connecting the two clusters. There is an interval of ε values in which there are still no cycles between the clusters, but the cycles within each cluster close up. This is the empty purple space between the flares. Once ε is large enough, cycles again form, now connecting the two clusters.

In addition to these two large clusters, we can see from the knots that we will have small clusters of points at each of the knot's local extrema. The number of local maxima is given by the n_z term in the parameterization. Knot (3, 2, 5, 1.5, 0.2) in Fig. 13, for example, has only five peaks, and the curvature at these peaks is less than that in other knots. As a result, the points in the mini-clusters at each local maximum and minimum are further spread apart, bringing these smaller clusters closer together. We see this by the wider and closer flares in Fig. 13 heatmap. On the other hand, knot (2, 3, 11, 0.2, 0.7) in Fig. 14 has eleven peaks and the points in the clusters at the peaks have a higher curvature than the other knots. The smaller clusters are thus denser, leading to larger gaps between these smaller clusters. The resulting heatmap for this knot has skinnier, more separated flares, as seen in Fig. 14.

The heatmaps we have studied thus far demonstrate that Lissajous knots with higher periodicity have skinnier flares and a larger range in curvature values that results in more open space at the top of the plot. This preliminary work suggests that this method may be useful in the study's other periodic time series datasets.

5 Classification of Anisotropic Kuramoto–Sivashinsky Solution

Indeed, there are many contexts where spatial structure of data is changing temporally. In this vein, we will explore a complex spatio-temporal pattern, using pseudo-multidimensional persistence to account for both the spatial and temporal variation of a pattern. Such patterns often occur in nonlinear systems that are driven from equilibrium by, for example, a gradient in temperature, concentration, or velocity [24]. Consider the coloration patterns of zebras or specific species of fish, ripples in sand dunes, or convection cell formations in clouds. Understanding these pattern-forming systems is important to a wide variety of fields in the scientific community such as biology, physics, engineering, and chemistry [7].

In modeling physical phenomena, it is often the case that poorly-resolved or poorly understood processes are parametrized rather than treated explicitly. Because of this, it becomes important to determine the influence of model parameters on the system. There are a variety of methods to do this, many of which require computationally expensive simulations [1]. Irregular time-varying structures and complexity of patterns, and sensitivity to initial conditions, among other things, make quantifying or even distinguishing patterns difficult [11]. Recently there has been much interest in using topological methods in pattern formation and pattern evolution, in particular in material sciences [28]. Computational topology has emerged as a tool that retains some essential information for studying patterns, while significantly reduces the dimensionality of the data [9]. For example, persistent homology has been used to distinguish between parameters for complex patterns formed through a phase separation process [11] in the Cahn–Hilliard equations. In this example, the patterns were studied at specific, static moments in time.

Pseudo-multidimensional persistence allows for the inclusion of the time evolution. We will apply this technique to simulations of the two-dimensional anisotropic Kuramoto–Sivashinsky equation.

The Kuramoto–Sivashinsky (KS) equation is a partial differential equation used to model systems driven from equilibrium [21]. It has found many applications in surface pattern-formation including flame front propagation [23], surface patterning by ion-beam erosion [8, 20], epitaxial growth and instabilities related to electromagnetism [27], the formation of suncups in snowfields [25], and solidification from a melt [15]. The solution $u(x, y, t)$ gives a patterned surface in two spatial variables that evolves in time. This equation arises in applications as surface nanopatterning by ion-beam erosion and solidification. The anisotropic Kuramoto–Sivashinsky (aKS) equation is given by

$$\frac{\partial u}{\partial t} = -\nabla^2 u - \nabla^2 \nabla^2 u + r \left(\frac{\partial u}{\partial x} \right)^2 + \left(\frac{\partial u}{\partial y} \right)^2, \tag{1}$$

where $\nabla^2 = \frac{\partial^2}{\partial x^2} + \frac{\partial^2}{\partial y^2}$. The parameter r controls the anisotropy in the nonlinear term. The goal of this experiment is to classify sets of solutions by parameter r. Numerical simulations of the aKS equation are generated for parameter values $r = 0.5, 0.75, 1, 1.25$, and 1.5. Thirty trials for each parameter value are generated using a low-amplitude white noise initial condition (Fig. 16). Persistent homology is computed using sublevel sets. At each threshold height, a cubical complex is built on neighboring points with values below the threshold. A filtration of cubical complexes is formed by increasing the threshold from below the surface to the maximum height of the surface. See [11] for an introduction to cubical homology [22].

This data set was initially investigated in the paper [1], with the goal of identifying parameter values of each example using persistent homology. Persistent homology was computed using a cubical complex on a sublevel set filtration at a single moment in time. In order to compare persistence diagrams, Adams et al. devised a stable vector representation of persistence diagrams called persistence images (PIs). Using this method, each persistence diagram is vectorized. Standard machine learning algorithms may then be applied to classify the vectors based on the parameter used to generate each surface. See the paper for full details on the method. Using a subspace discriminant ensemble, which uses the same classification algorithm repeatedly over randomly chosen subspaces of the data. The algorithm fits the data by building a model on the mean and variance of classes in a training set. Each example in the testing set is assigned a class for each iteration. A likely overall class is assigned at the end. In this example, the data was classified at several different time steps. The classification accuracies reported in their experiment are listed in Table 3. Using this method, one must choose a single moment in time to consider the data. We see the restriction to a single moment in time as a limiting factor because the temporal evolution of the pattern is ignored. The vectors representing several time steps may be concatenated into a larger vector to be used for classification, but this causes the size of the vectors to grow quickly,

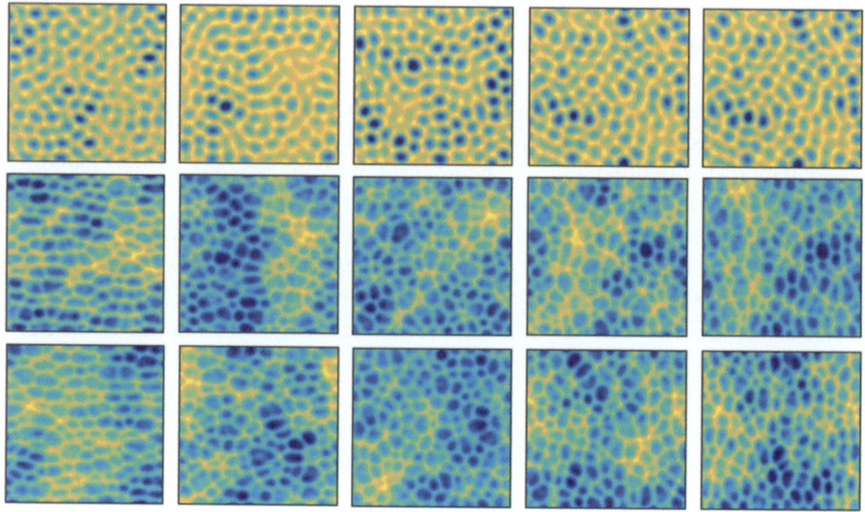

Fig. 16 Plots of the examples of the surfaces $u(x, y, \cdots)$ from the numerical simulations. Each column is generated using a different parameter value; $r = 0.5, 0.75, 1.0, 1.25$, and 1.5. Each row is a single point in time in the evolution of the pattern of the surface, the times shown are $t = 3, 5$ and 10. By $t = 5$ the elongation due to the anisotropy has stabilized, though the surface continues to evolve in time. $r=1$ (center column) is the isotropic case; there is no elongation in either direction

Table 3 Classification accuracies at different times of the aKS solution, using a fivefold cross-validated subspace discriminant ensemble on a vector representation of one-dimensional barcodes

Persistence representation	Time t=3 (%)	Time t=5 (%)	Time t=10 (%)
β_0 PIs	58.3	96.0	94.7
β_1 PIs	67.7	87.3	93.3
β_0 and β_1 PIs	72.7	95.3	97.3

Classification of times $t = 15$ and 20 results in accuracies similar to $t = 10$

which limits the machine learning techniques that may be efficiently applied. Even with a low resolution, generating persistence images at each time step results in a high-dimensional representation of the data. We consider heatmaps as an alternative representation to capture the evolution of the pattern over time in a way that lends itself to machine learning techniques.

A heatmap will be generated for each example. One dimension will be the height of the sublevel set filtration, represented along the horizontal axis, and the other dimension will be the time step, represented along the vertical axis. The top row represents the earliest time step. Each entry is given by $ln(\beta_i + 1)$ where β_i is the Betti number in dimension 0 or 1. A single heatmap then will give the homological feature information of the surface as it evolves in time. Examples of the heatmap in each homological dimension are shown in Fig. 17. The earliest time is shown on the top row. The pattern is just beginning to form at this stage, so topological features

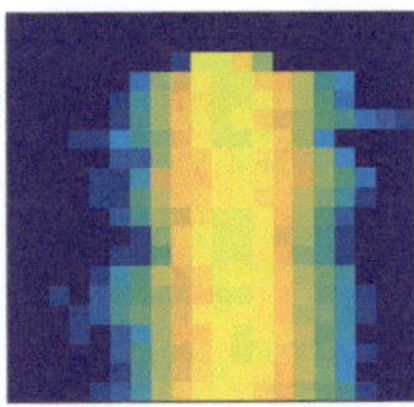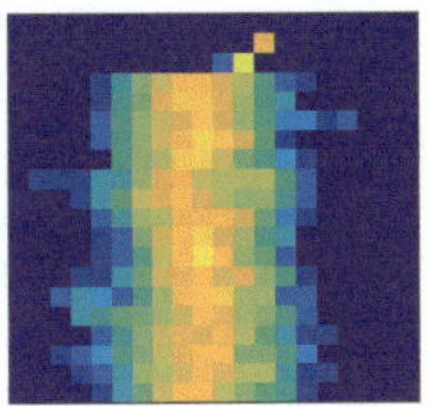

Fig. 17 Examples of heatmaps for the aKS equation, $r = 1$. The left shows zero-dimensional homology and the right is one-dimensional homology. The top row is the earliest time step. The pattern takes some time to emerge, so there in little by way of topological features at the early times

Table 4 Classification accuracies using the MDS reduction of heatmaps and the MDS reduction of PIs concatenated to incorporate several time steps

Homological dimension	Accuracy of MDS reduced heatmap (%)	Accuracy of MDS reduced PIs, t=5,10,15 (%)
β_0	69.3	72.0
β_1	100	98.7
β_0 and β_1	100	100

MDS reduces the overall dimension to 2. Classification was performed using a fivefold cross-validated linear discriminant

are still emerging as well. Interestingly, even though the pattern is dynamically changing, there are not significant changes from one row to the next for much of the time represented here. Once the heatmaps are formed, the L_2 distance is computed between each heatmap. As a natural dimension reduction step, we perform MDS with a chosen dimension of 2. At two-dimensions, 95% of the variance of the data is accounted for. The MDS representation is then used for classification. Similar to [1], classification is performed with a linear discriminant ensemble with fivefold cross-validation. In this case, there is no need to incorporate an ensemble of random subspaces since the dimension of the data is 2. The classification accuracy of the two-dimensional MDS for the heatmaps from β_0 and β_1 and the concatenated information are given in Table 4. We include a comparison with classification accuracies of PIs that have been concatenated to include several time steps, t=5,10, and 15. The PIs were also reduced to 2 dimensions with MDS. Classification accuracy is comparable when using the full time evolution (heatmaps), or several discrete time steps (PIs). It is clear that β_1 is much more discriminating than β_0. This is due to the nature of the pattern, which appears as raised bubbles. In a sublevel set filtration, β_1 will capture these features. Good classification results for the full time evolution are encouraging because it suggests that heatmaps will prove useful to capture more complete information on temporal evolution. In this case, enough information was captured in several time steps that a method such as PIs could be used, but in general this may not contain enough information to discriminate between classes. Heatmaps allow users to make use of more temporal data that may be available to them. Heatmaps (combined with a dimension reduction technique) provided very low dimensional representations of dynamically changing patterns, that allowed for good parameter recovery.

6 Conclusion and Future Research

In this article, we proposed the heatmap as a tool to approximate multidimensional persistent homology and explored applications to a variety of topics, including shape analysis in 3D craniofacial imaging, Lissajous knots, and the solutions of the anisotropic Kuramoto–Sivashinsky equation. Our use of heatmaps in these areas allowed the incorporation of additional parameters such as scale, curvature, and time, into the regular univariate filtrations of persistent homology. The addition of a second parameter, particularly in our work with the Kuramoto–Sivashinsky equation, allowed a new avenue of insight into the datasets studied that are inherently described by multiple parameters.

As demonstrated in this paper, this technique for the approximation of multi-dimensional persistence has potential for a wide range of application in numerous fields. However, the techniques presented here are certainly still in their infancy and there remains much work to be done in the future to make them competitive with existing methods, particularly in terms of computational time and classification accuracy. These issues, including the determination of a dissimilarity appropriate for comparing heatmaps as well as the excessive computational time required by our own and other existing techniques for topological data analysis, will need to be addressed before these methods become widely applicable.

Acknowledgements We thank the Workshop for Women in Computational Topology (WinComp-Top) held on August 12–19, 2016, Institute for Mathematics and its Applications (IMA). We would like to thank the Seed Grant from Women and Children's Health Research Institute, the National Sciences and Engineering Research Council of Canada (NSERC), the McIntyre Memorial fund from the School of Dentistry at the University of Alberta, and Biomedical Research Award from American Association of Orthodontists Foundation. We are grateful for sleep scholars, research coordinators, and postdoctoral fellows, and technicians in Stollery Children's Hospital, University of Alberta.

References

1. H. Adams, S. Chepushtanova, T. Emerson, E.M. Hanson, M. Kirby, F.C. Motta, R. Neville, C. Peterson, P.D. Shipman, L. Ziegelmeier, Persistence images: a stable vector representation of persistent homology. J. Mach. Learn. Res. **18**, 1–35 (2017)
2. A. Al-Sharadqah, N. Chernov, Error analysis for circle-fitting algorithms. Electron. J. Stat. **3**, 886–911 (2009)

3. M.G.V. Bogle, J.E. Hearst, V.F.R. Jones, L. Stoilov, Lissajous knots. J. Knot Theor. Ramif. **3**(02), 121–140 (1994)
4. P. Bubenik, Statistical topological data analysis using persistence landscapes. J. Mach. Learn. Res. **16**, 77–102 (2015)
5. G. Carlsson, A. Zomorodian, The theory of multidimensional persistence. Discrete Comput. Geom. **42**(1), 71–271 (2009)
6. A. Collins, A. Zomorodian, G. Carlsson, L.J. Guibas, A barcode shape descriptor for curve point cloud data. Comput. Graph. **28**, 881–894 (2004)
7. M. Cross, H. Greenside, *Pattern Formation and Dynamics in Nonequilibrium Systems* (Cambridge University Press, Cambridge, 2009)
8. R. Cuerno, A.-L. Barabási, Dynamic scaling of ion-sputtered surfaces. Phys. Rev. Lett. **74** 4746 (1995)
9. S. Day, W.D. Kalies, T. Wanner, Verified homology computations for nodal domains. Multiscale Model Simul. **7**(4), 1695–1726 (2009)
10. T.K. Dey, J. Sun, Y. Wang, Approximating cycles in a shortest basis of the first homology group from point data. Inverse Prob. **27**, 124004 (2011). An International Journal on the Theory and Practice of Inverse Problems, Inverse Methods and Computerized Inversion of Data
11. P. Dłotko, T. Wanner, Topological microstructure analysis using persistence landscapes. Phys. D **336**, 60–81 (2016)
12. H. Edelsbrunner, J. Harer, *Computational Topology: An Introduction* (American Mathematical Society, Providence, 2010)
13. C. Flores-Mir, M. Korayem, G. Heo, M. Witmans, M.P. Major, P.W. Major, Craniofacial morphological characteristics in children with obstructive sleep apnea syndrome: a systematic review and meta-analysis. J. Am. Dent. Assoc. **2013**, 269–277 (2013)
14. R. Ghrist, *Elementary and Applied Topology* (Createspace Independent Pub, Seattle, 2014)
15. A.A. Golovin, S.H. Davis, Effect of anisotropy on morphological instability in the freezing of a hypercooled melt. Phys. D **116** 363–391 (1998)
16. S. Kullback, R.A. Leibler, On information and sufficiency. Ann. Math. Stat. **22**, 79–86 (1951)
17. H. Lee, H. Kang, M.K. Chung, S. Lim, B.N. Kim, D.S. Lee, Integrated multimodal network approach to PET and MRI based on multidimensional persistent homology. Hum. Brain Mapp. **38**, 3871402 (2017)
18. M. Lesnick, M. Wright, Interactive visualization of 2-D persistence modules (2015). arXiv:1512.00180
19. C.L. Marcus, L.J. Brooks, S.D. Ward, K.A. Draper, D. Gozal, A.C. Halbower, J. Jones, C. Lehmann, M.S. Schechter, S. Sheldon, R.N. Shiffman, K. Spruyt, Diagnosis and management of childhood obstructive sleep apnea syndrome. Pediatrics **130**, 576–584 (2012)
20. F.C. Motta, P.D. Shipman, R.M. Bradley, Highly ordered nanoscale surface ripples produced by ion bombardment of binary compounds. J. Phys. D Appl. Phys. **45**(12), 122001 (2012)
21. D. Papageorgiou, Y. Smyrlis, The route to chaos for the Kuramoto-Sivashinsky equation. Theor. Comput. Fluid Dyn. **3**(1), 15–42 (1991)
22. D. Rolfsen, *Knots and Links*. Sistema Librum 2.0 (SERBIULA, 2018), pp. 391–415
23. G.I. Sivashinsky, Instabilities, pattern formation, and turbulence in flames. Annu. Rev. Fluid Mech. **15**(1) 179–199 (1983)
24. H.L. Swinney, Emergence and evolution of patterns, in *Proceedings of AIP* (1999), pp. 3–22
25. T. Tiedje, K.A. Mitchell, B. Lau, A. Ballestad, E. Nodwell, Radiation transport model for ablation hollows on snowfields. J. Geophys. Res. **11** (F2) 2156–2202 (2006)
26. C.M. Topaz, L. Ziegelmeier, T. Halversion, Topological data analysis of biological aggregation models. PLoS ONE **10**(5), e0126383 (2015)
27. J. Villain, Continuum models of crystal growth from atomic beams with and without desorption. J. Phys. I Fr. **1** 19–42 (1991)
28. T. Wanner, Topological analysis of the diblock copolymer equation, in *Mathematical Challenges in a New Phase of Materials Science* (Springer, Berlin, 2016), pp. 27–51

29. K. Xia, G.-W. Wei, Multidimensional persistence in biomolecular data. J. Comput. Chem. **36**, 1502–1520 (2015)
30. M. Younes, W. Thompson, C. Leslie, T. Egan, E. Giannouli, Utility of technologist editing of polysomnography scoring performed by a validated automatic system. Ann. Am. Thorac. Soc. **12**, 1206–18 (2015)